Modern Appli Fracture Mechanics

Modern Applied Fracture Mechanics

Cameron Coates

Valmiki Sooklal

CRC Press
Taylor & Francis Group
Boca Raton London New York

CRC Press is an imprint of the
Taylor & Francis Group, an **informa** business

First edition published 2022
by CRC Press
6000 Broken Sound Parkway NW, Suite 300, Boca Raton, FL 33487-2742

and by CRC Press
4 Park Square, Milton Park, Abingdon, Oxon, OX14 4RN

CRC Press is an imprint of Taylor & Francis Group, LLC

© 2022 Taylor & Francis Group, LLC

Library of Congress Cataloging-in-Publication Data
Names: Coates, Cameron, author. | Sooklal, Valmiki, author.
Title: Modern applied fracture mechanics / Cameron Coates, Valmiki Sooklal.
Description: First edition. | Boca Raton : CRC Press, [2022] | Includes bibliographical references and index. |
Identifiers: LCCN 2021055513 (print) | LCCN 2021055514 (ebook) | ISBN 9780367501259 (hardback) | ISBN 9780367509880 (paperback) | ISBN 9781003052050 (ebook) Subjects: LCSH: Fracture mechanics.
Classification: LCC TA409 .C63 2022 (print) | LCC TA409 (ebook) | DDC 620.1/126--dc23/eng/20220105 LC record available at https://lccn.loc.gov/2021055513LC ebook record available at https://lccn.loc.gov/2021055514

ISBN: 978-0-367-50125-9 (hbk)
ISBN: 978-0-367-50988-0 (pbk)
ISBN: 978-1-003-05205-0 (ebk)

DOI: 10.1201/9781003052050

Typeset in Times
by SPi Technologies India Pvt Ltd (Straive)

Dedication

To June, Nia, Kimani, Naomi, late brother Mark, sister Camille, and my parents Clinton (late) and Ava Coates of Ensom City, Jamaica, for their encouragement, faith, and support.

Cameron W. Coates

To my wife Tina and my parents Donna and Karamchand of Arouca, Trinidad and Tobago, for their love and support in all my endeavors.

Val Sooklal

Contents

Preface

Fracture Mechanics (FM) is a relatively new field (beginning in the 1940s) that has made a tremendous impact on our ability to predict and prevent structural failure. It will become increasingly important in the 21st century as scientists and engineers strive to develop custom-engineered materials at the nano-, meso-, and macro-levels, as well as fabrication processes that are more capable, sustainable, portable, and flexible. Additionally rapid advancement in processing speeds and computational methods will allow more efficient, programmable, and scalable FM applications.

This text is designed to provide undergraduate engineers with a basic introduction to Linear Elastic Fracture Mechanics (LEFM) with an emphasis on modern applications, including current software elementary applications. Undergraduate engineering and science students who expect to work in areas in which critical structural or material failures occur – such as aerospace, petroleum, production, civil, biomedical, food processing (equipment), nuclear, mechatronics, mechanical, or manufacturing – will find the analytical tools and software process examples in this text particularly valuable. Practicing engineers, scientists, or technicians who were not exposed to FM as undergraduates can also utilize this text as a nuts-and-bolts resource for common approaches, and as an aid in developing a solid understanding of FM fundamentals with minimal mathematical complexity.

FM is more commonly taught at the graduate level for civil, mechanical, aerospace, materials science and manufacturing engineers; however, an increasing diversity of new materials as well as interdisciplinary design efforts will require broader everyday use of these techniques, even for simple designs that have traditionally been strength-based. It is therefore not an unreasonable expectation that undergraduate engineers of all types will need to take more advanced coursework in this area, beyond the introductory content taught on a standard undergraduate materials science and engineering course. Proper use of this text will enable the graduate engineer on transition into a professional field that utilizes FM tools to efficiently interpret and apply practical LEFM solutions, to easily adapt to and understand how to use FM software, and to conduct or develop LEFM related experiments.

Ideally, the strength-based techniques taught on a traditional undergraduate Mechanics of Materials course should be studied prior to covering the material in this text. There is relatively little calculus in this text; however, knowledge of differential and integral calculus, standard in an undergraduate engineering curriculum, is assumed in a few theoretical formulations. Several FM texts already exist; however, the majority contain material taught at the graduate level with complex mathematics and limited modern day applications. While a thorough understanding of the complex mathematics is necessary for engineers to advance current theories, to develop specialized software or to solve uniquely complex FM problems, many routine applications require less mathematical rigor but greater practical knowledge, particularly as it pertains to using software-based solutions.

Several software packages exist which allow engineers to apply linear and nonlinear FM methods to evaluate structural integrity. Their usage and capability can only be expected to increase as technology marches forward. It is now more important than ever that all engineering undergraduates who might work with structures and materials are provided with the fundamental tools needed to properly input FM information in the various FM software packages that currently exist (e.g. AFGROW, NASGRO, ABAQUS, ANSYS) or will exist in the future. This text offers these tools by providing industry applications and case studies, and by discussing the model choice, assumptions, and common language used in FM software. The authors would like to thank Dr. Valeria La Saponara, Mr. James Harter (AFGROW), and Dr. Ashok Saxena for reviewing some aspects of this text. We also thank graduate student Shyam Patapati, undergraduates Austin Webb, Jose Bonilla-Martinez, and Ryan Foster for assistance with figures and providing a student perspective regarding the material.

Authors

Cameron Coates is Professor of Aerospace Engineering and Assistant Dean at Kennesaw State University, in Marietta, Georgia. Dr. Coates earned his B.Sci from the United States Naval Academy, and his M.Sci and Ph.D from Georgia Tech University. His specialties include solid mechanics, fracture and fatigue analysis, smart structures and engineering education; he is active in research and service-based activities aimed at improving student success.

Valmiki Sooklal is Associate Professor of Mechanical Engineering at Kennesaw State University. His research areas include experimental and computational fracture mechanics, laser/material interaction, tissue welding, sustainable housing, and remote sensing. Dr. Sooklal graduated from Tulane University, in New Orleans, with a doctorate in Mechanical Engineering. He also conducts research in Engineering Education, with a focus on developing education tools to enhance the senior design/capstone process.

1 Fracture Mechanics

OBJECTIVES

After studying this chapter, the student should be able to:

1. Explain the need for Fracture Mechanics using historical cases.
2. Explain the differences between a Fracture Mechanics-based approach and a Strengths of Materials approach.
3. Understand and explain the state of stress at a point, plane stress, and plane strain for a body under applied loads.
4. Describe the typical fracture behavior of metals, polymers, and ceramics.

1.1 HISTORICAL TO RECENT FAILURES

Engineers have wrestled with failure prevention since the first man-made structure was constructed. Many of our most devastating structural failures since the early 20th century have been attributed to a lack of understanding of the mechanics of fracture. These failures have been widespread among diverse engineering areas, including but not limited to: pressure vessels in the nuclear, food processing, petroleum, and space industries; aircraft structures and components; buildings; bridges; ships; automobile structures and components; as well as biomedical implants. Only a few structural failures due to fracture will be highlighted below as there are several books and websites that discuss these failures in greater detail.

During World War II, the allied force ships that were used to transport ammunitions, supplies, and equipment were continuously lost to German U-boats and submarines. This led to the adoption of an emergency ship-building program focused on rapid construction. In response to the demands for rapid ship construction, some 2,500 Liberty ships were constructed with a new design paradigm; primarily, welds were used for all external structural connections. Prior to this, ship construction utilized both welds and rivets for joining metal plates. Within two years of service, ten Liberty ships had major fractures in their hulls. Over the course of the war, some 1,200 ships suffered from cracks and three were lost by suddenly splitting in two. A Liberty ship fractured from deck to keel is shown in Figure 1.1(a).

A Board of Investigation established by the United States Marine Corps concluded that defective workmanship (particularly for the weld seams) was an identifiable contributing factor in half of the fracture incidents [1]. Other contributing factors included: the quality of the steel, the use of unskilled workers, inadequate supervision, and poor design details. The board also funded 30 distinct research projects at laboratories throughout the United States [2]. Dr. Constance Tipper, an engineering professor at Cambridge University in England, found that the grade of steel used to make the Liberty ships was particularly vulnerable to embrittlement when subjected to low temperatures in the North Atlantic. She concluded that

DOI: 10.1201/9781003052050-1

(a)

(b)

FIGURE 1.1   (a) Liberty Ship *Schenectady* in the port of Portland, fractured from deck to keel [5] (b) Constance Tipper (1894–1995), pioneer in the field of the brittle/ductile fracture of metals.

existing flaws were able to grow more rapidly when the material was brittle and that the cracks could propagate across very large distances due to the welds. During the investigation of the Liberty ships, Dr. Tipper developed a test that became the standard method for determining brittleness in steel – now commonly known as the "Tipper Test." This test helps to predict whether a type of steel is more likely to behave in a ductile or brittle manner at service temperatures [3]. Constance Tipper, shown in Figure 1.1(b), was a pioneer regarding advancements in the study of ductile/brittle fracture in the 1950s.

The De Havilland Comet I shown in Figure 1.2 was the world's first pressurized commercial jet airliner. It was launched into service in 1952; unfortunately, three Comets failed catastrophically in service. All three broke apart midflight killing all on board. Several others had substantial structural failures that led to the grounding of the entire fleet of aircraft. Sir Arnold Hall [4], director of the Royal Aircraft Establishment, led a team of engineers and scientists in the investigation of the cause of these failures. The team found that the cracks had initiated from the corners of the doors and windows of the aircraft due to the pressurization–depressurization cycles during service. The square window geometry introduced stress concentrations at the corners, resulting in crack initiation. The specific riveted construction technique used was also found to result in the service tensile stresses exceeding their design capacity.

The Aloha Airlines 1988 accident [6] involved the total separation of the upper aircraft skin and fuselage while in flight, which resulted in one fatality and eight serious injuries. This accident spurred advances in damage tolerant design methods, particularly in the case of multiple site damage in aircraft structures. "Damage tolerance" refers to the ability of structural components to perform despite containing cracks that may or may not be growing.

On October 28, 2016, American Airlines flight 383, a Boeing 767-323, N345AN, underwent an uncontained engine failure in the right engine and a subsequent fire

FIGURE 1.2   De Havilland DH106 Comet 1.

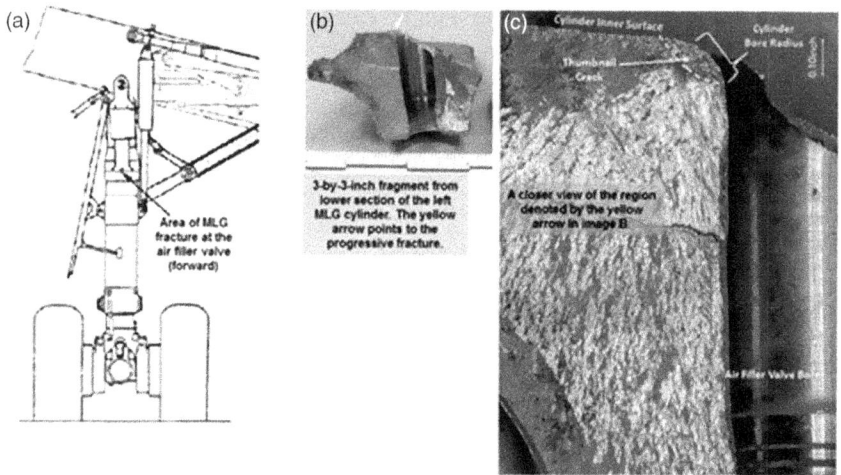

FIGURE 1.3   (a) Diagram of an MD-10-10F left MLG, (b) photograph of the three-by-three-inch fragment from the lower section of the left MLG cylinder, (c) closer view of the crack initiation region in image [8].

during its takeoff ground roll at Chicago O'Hare International Airport (ORD), Chicago, Illinois. The National Transportation and Safety Board (NTSB) concluded that the right engine experienced an uncontained high-pressure turbine (HPT) stage two disk rupture during the takeoff roll. The NTSB further concluded that the failure was due to multiple low-cycle fatigue cracks that initiated from an internal material anomaly, known as a discrete dirty white spot, which formed during the processing of the disk material [7]. When a cyclical load is applied to a component, the resulting cyclical stress fields may induce crack nucleation and growth despite maximum stress values being well below the material's failure stresses; these resulting cracks are called fatigue cracks.

Crack growth may of course be exacerbated by environmental effects such as moisture and oxidation, that is corrosion. On October 26, 2016, the main landing gear (MLG) of a FedEx MD-10-10F broke during landing at Fort Lauderdale/ Hollywood International Airport. This led the plane to yaw to the left and resulted in a fuel-fed fire on the left wing. The MLG failure was caused by corrosion that led to fatigue (cyclical applied loads) cracking. Once the crack had progressed to a critical length, the left MLG cylinder fractured in overstress due to the loads imposed during landing [8]. The MLG fractured region is shown in Figure 1.3. A fracture surface that has occurred because of fatigue loads yields distinct patterns called striations. These striations can be used to determine the number of cycles that occurred prior to failure. This application will be discussed in detail in Chapter 4.

As we continue to develop advanced materials, the diversity of failure mechanisms and modalities will increase. The field of fracture mechanics is relatively new (since 1948) and can be expected to evolve at a rapid pace. Therefore, some of the original theories discussed in this text will likely be modified and new theories advanced as we develop new materials and methods of manufacturing.

1.2 THE NEED FOR FRACTURE MECHANICS AND THEIR APPLICATIONS

Regardless of fabrication method, all materials will have imperfections at both the microscopic and macroscopic level. Microscopic imperfections include interstitials, vacancies, impurities, dislocations, and microcracks. It is these imperfections that may serve as crack nucleation sites. Additionally, general usage and maintenance events can also result in the nucleation of microcracks.

1.3 MATERIALS SCIENCE REVIEW

For certain materials, upon solidification, the atoms will position themselves in a repetitive three-dimensional (3D) pattern over large atomic distances. A single array of this repeated 3D structure is called a crystal, and the array will have a specific orientation. This class of materials are referred to as "crystalline." As a crystalline material solidifies several crystals will form independently with different orientations. The boundary where these crystals meet is called the "grain boundary," as shown schematically in Figure 1.4. The grain boundary as well as the way the atoms, ions, or molecules are spatially arranged influences the fracture toughness. All metals, certain polymers, and ceramics exhibit crystallinity.

1.4 DISLOCATIONS AND PLASTICITY

A dislocation is a 1D defect within a crystalline structure. One such defect, referred to as an edge dislocation, occurs if a plane of atoms within the structure is discontinuous and the edge of the plane, called the dislocation line, lies within the crystal. Atomic distortions occur in the vicinity of the dislocation line. In another case, when a segment of the crystalline structure is subjected to shear loading, the segment may displace, breaking atomic bonds and forming new ones. The planes of atoms will then trace a spiral or helical path around the dislocation line; this is called a screw dislocation. "Plasticity" or "plastic deformation" refers to the phenomenon

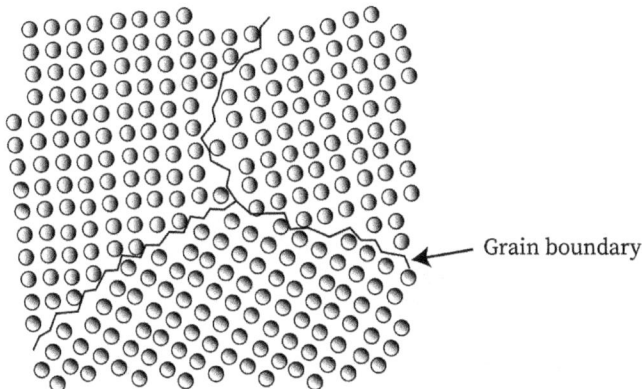

FIGURE 1.4   Schematic diagram of grain boundary.

in which a solid material undergoes permanent deformation. Dislocations are the primary drivers of plastic deformation, and both types of dislocations result from shear loading.

1.5   ISOTROPY VS. ANISOTROPY

The material properties (e.g. Young's modulus, Poisson's ratio) for isotropic bodies are uniform regardless of orientation, that is these properties are independent of direction. Therefore, these materials will have the same response if the same stress (due to an external load) is applied in a different direction. Metals and most thermoplastics are isotropic. Anisotropic materials have directionally dependent properties and will respond differently if the same stress is applied in a different direction. Though a material is isotropic, components made with that material do not necessarily retain isotropy. Composites and 3D-printed polymers are anisotropic. If the service loading is known a priori, the application of anisotropy can be an engineering decision. For example, if only tensile loading is expected, a fiber-reinforced composite may be used where fibers are unidirectional and aligned with the expected tensile loads.

Concept Challenge 1.1

Can a material with anisotropic regions be itself isotropic?

1.6   STATE OF STRESS AND STRAIN CONCEPTS

In solid mechanics, the state of stress at a point is defined using a cubic element in equilibrium within a Cartesian coordinate system, with its origin at the center of the cube. For the most general loading case, each outward facing plane is subjected to a normal stress, that is the stress acting perpendicular to the face, and two mutually orthogonal shear stresses, defined as acting along the plane. This is shown in Figure 1.5.

Let us define the plane by the axis direction to which it is perpendicular; therefore the plane perpendicular to the x-axis is denoted by x. We can then define each stress acting on a plane using double subscripts; the first subscript represents the outward facing plane on which the stress acts and the second subscript represents the axis direction in which the stress acts (see Figure 1.5). For example, σ_{xx}, a normal stress, acts on outward facing planes that are perpendicular to the x-axis and in a direction parallel to the x-axis; σ_{xy}, a shear stress, acts on the face perpendicular to the x-axis and in a direction parallel to the y-axis. Note, in this text σ_{ij} ($i \neq j$) is used to represent shear stress; many authors also use $\tau_{ij}(i \neq j)$. Enforcement of force equilibrium in each Cartesian direction will require that normal stresses on opposite faces are equal and opposite. The enforcement of moment equilibrium will require that shear stresses acting on adjacent perpendicular faces in directions that are coplanar (or on parallel planes) be equal, that is $\sigma_{xy} = \sigma_{yx}$, $\sigma_{xz} = \sigma_{zx}$, and $\sigma_{yz} = \sigma_{zy}$.

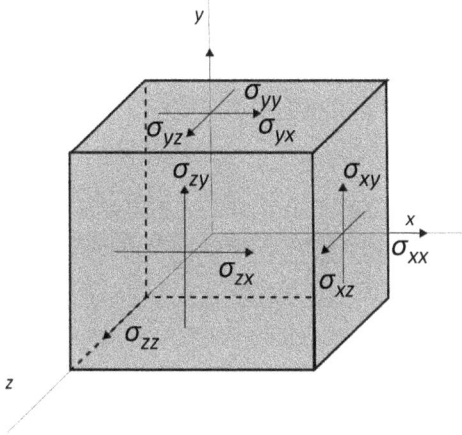

FIGURE 1.5   State of stress for a cubic element.

We can represent all nine stresses using a single expression, σ_{ij}, where i represents the face on which the stress is acting (the face is designated by the axis which it is perpendicular to) and j represents the axis direction in which the stress acts.

Let i and j represent the x, y, or z directions. We can then write the state of stress using a 3×3 matrix as shown below.

$$\sigma_{ij} = \begin{pmatrix} \sigma_{xx} & \sigma_{xy} & \sigma_{xz} \\ \sigma_{yx} & \sigma_{yy} & \sigma_{yz} \\ \sigma_{zx} & \sigma_{zy} & \sigma_{zz} \end{pmatrix} \tag{1.1}$$

Note that stress is NOT a vector, it is a second-order tensor. The term "tensor" describes any physical quantity that can be represented by a scalar, vector, or a matrix. Physical properties such as mass, temperature, or density, which are scalars, would be considered zero-order tensors. Physical properties that may be represented as vectors such as force, velocity, or acceleration are first-order tensors. Stress is a second-order tensor as the stress state at a point is represented by a matrix. Moment equilibrium requires that $\sigma_{ij} = \sigma_{ji}$ $(i \neq j)$, therefore the six independent shear stress components reduce to three. The total number of independent stress components will therefore be six; three independent shear stresses and three independent normal stresses.

1.7   STRESS-BASED APPROACH

In the traditional stress-based approach, we design structures by ensuring that in-service stresses or loads do not exceed certain maximum values dictated by the material mechanical properties. These maximum values are based on computational modeling predictions, experimental results, analytical predictions, or a combination of the three approaches. In the design approach model, the material is typically assumed to be linear elastic, isotropic, and a continuum, that is no flaws exist within the material. Inevitably, some safety factor, SF, is chosen that satisfies:

$$SF = \frac{\sigma_m}{\sigma_a} \qquad (1.2)$$

σ_m will represent the magnitude of a specific type of stress (e.g. tensile) at which the material yields or fractures; this stress value is typically referred to as "strength," while σ_a will represent the same type of stress expected in service, sometimes referred to as the "allowable stress." If the material is, for example, only loaded in a tensile manner during service, then only tensile stresses need be considered. In this case, Eqn. (1.2) would be all that is needed. However, most load applications are a combination of axial (tension and compression), lateral, and torsional loads. In these cases, an element within the material will be subjected to shear and normal stresses. Additionally, a variety of maximum stresses may have to be considered, for example tensile, compressive, and torsional. The safety factor accounts for: the probability of overload, whether the loads are static or dynamic (cyclical vs. random); the construction or fabrication technique; deterioration due to environmental effects; variability in workmanship; assumptions in the theoretical methods used; and the failure consequences. The choice of the magnitude of SF is a complex one, usually established by a collaboration of experienced engineers. Many codes have been developed by professional engineering organizations such as the American Society of Mechanical Engineers (ASME), which provide guidelines or recommendations for the SF for various design scenarios.

Example 1.1

A cylindrical aluminum bar (Al 2024-T4) with a diameter of 20 mm is subjected to an axial tensile load of 29,846 N. Given that the tensile yield strength, σ_{ys}, of Al 2024-T4 is 276 MPa, determine whether the bar remains within its elastic limit under the applied load. If this bar is not to exceed its elastic limit during use and the generally accepted Safety Factor is 3, would you recommend that AL 2024-T4 be used in this case?

SOLUTION

The "yield strength" is the stress value at which aluminum's response to mechanical stimuli transitions from elastic to plastic; this number is typically obtained from experimental data. We can first determine the normal stress in the bar due to the applied load, and then compare this value to the yield strength. If the normal stress exceeds the yield strength, then we know the bar has exceeded its elastic limit.

Given a cross-sectional area, A, and an applied load, P, the normal stress, σ, is determined from

$$\sigma = \frac{P}{A}$$

Therefore, the normal stress,

$$\sigma = \frac{29{,}846\,\text{N}}{\pi \times (0.01\,\text{m})^2} = 9.5 \times 10^7 \,\text{N/m}^2 = 95\,\text{MPa}$$

The normal stress of 95 MPa due to the applied load is substantially less than the yield strength (276 MPa), therefore we may conclude that the bar elongates but remains within its elastic limit.

If, however, we incorporate a Safety Factor of 3, then

$$3 = \frac{\sigma_{YS}}{\sigma_a} = \frac{276\,\text{MPa}}{\sigma_a}$$

$$\Rightarrow \sigma_a = 92\,\text{MPa}$$

Since the applied normal stress of 95 MPa exceeds the allowable stress of 92 MPa, this material would not be recommended for this bar.

1.8   THE FRACTURE MECHANICS APPROACH

A flaw may be defined as a disruption in material homogeneity, such as a crack, a void, or an undesirable foreign inclusion. The presence of this flaw alters the local stresses such that the stress-based approach (linear elastic continuum) may predict a failure stress substantially above the stress magnitude that results in failure. When flaw sizes and orientations are considered, failure locations may also be different from those predicted using the stress-based approach. So unfortunately, in many situations the stress-based approach to structural analysis is insufficient. The fracture mechanics approach accounts for the effect of flaw size, location, orientation, and the number of flaws in the determination of failure stress. Fracture mechanics generates criteria for failure based on the interaction of these flaw characteristics with applied stresses and material and geometric properties.

In the general fracture mechanics approach, a property that provides a measure of the material's resistance to brittle fracture, called "fracture toughness," is used in the same manner that "strength" is used in the stress-based approach. There are two fundamental approaches in fracture mechanics: one incorporates a measure of stress intensity while the other uses an energy criterion. The stress intensity approach utilizes a parameter, K, called the stress intensity factor. In order to avoid failure due to fracture, K must remain below some critical value, K_C. The critical value is also called the "fracture toughness." This is a material property, similar to yield strength, elastic modulus, and so on. Linear elastic fracture mechanics (LEFM) will be addressed in detail in Chapter 2. The Safety Factor from a fracture mechanics approach would then be given by

$$SF = \frac{K_c}{K} \tag{1.3}$$

Note Eqn. (1.3) would be true for LEFM only, defined in Section 1.9. There is also an energy criterion that states that the energy per unit crack length available for crack growth must attain some maximum value for this growth to occur. The energy-based approaches will be discussed in detail in Chapter 3.

1.9   LINEAR ELASTIC VS. ELASTIC PLASTIC FRACTURE MECHANICS

If an applied stress is linearly proportional to the resulting stress for a given material up to the yield point, the material is defined as linear elastic. The linear elastic material response is shown in Figure 1.6(a). Consider an edge crack within a thin-walled plate

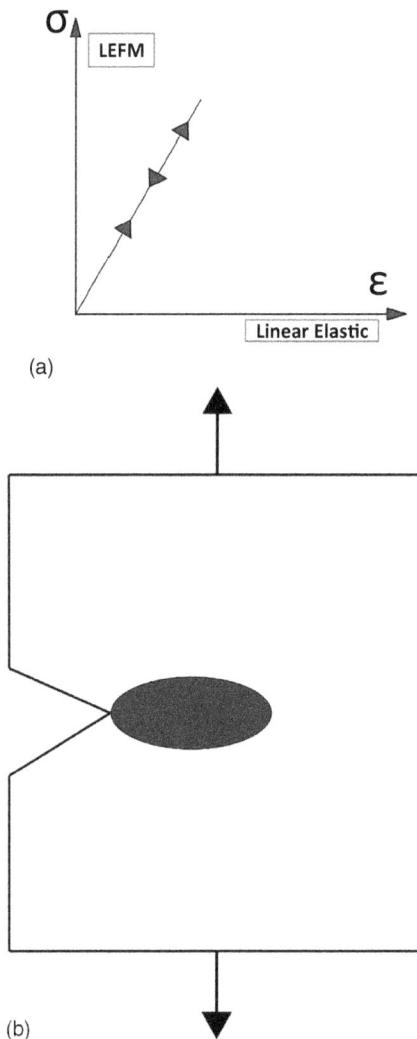

FIGURE 1.6   (a) Stress strain curve for linear elastic material. (b) Plastic zone at crack tip for edge crack in thin plate under tension.

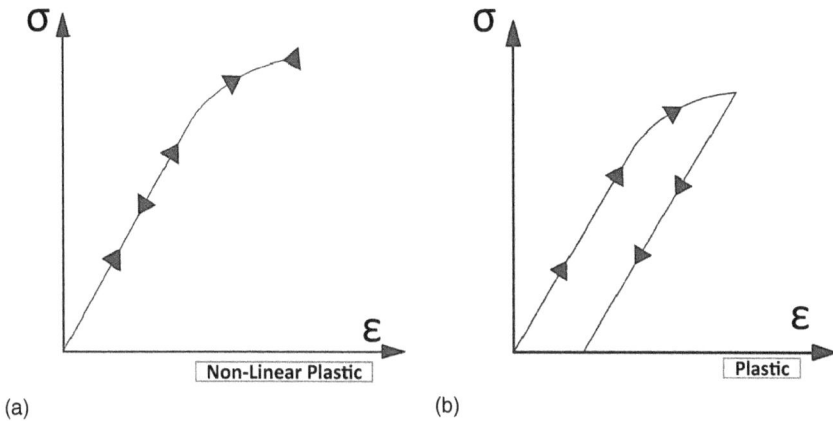

FIGURE 1.7   (a) Stress strain curve for nonlinear elastic material. (b) Stress strain curve for elastic-plastic material.

made of linear elastic material that is subject to tension, as shown in Figure 1.6(b). The stresses around the crack tip will increase beyond the yield strength of the material (this increase will be explained in subsequent sections), thereby generating a local plastic zone. LEFM is typically valid when this plastic zone is confined to a small region; this is usually true for brittle materials.

When elastic plastic (ductile) materials are subjected to loading, nonlinear deformation will occur at much lower applied loads compared to linear elastic materials. Therefore, nonlinear deformation is not confined to a small local area. Additionally, the deformation behavior may be time dependent. Elastic-plastic fracture mechanics (EPFM) is used to predict the fracture behavior of nonlinear elastic isotropic materials. For these materials, the unloading curve follows the original loading curve as shown in Figure 1.7(a). True elastic-plastic materials follow the unloading curve in Figure 1.7(b). In EPFM, either the strain energy fields or the displacement between the crack faces as the crack grows are used to describe the conditions at the crack tip and to establish a fracture criterion.

1.10   FRACTURE OF METALS

Fracture is defined as the separation of the material due to an applied stress, at a temperature below its melting point. Fracture is generally divided into two stages: crack initiation and crack propagation. Fracture characteristics will depend on the applied load (static, dynamic), temperature, and the type of material (metal, ceramic, polymer, composite). For metals, typically fracture is classified as brittle or ductile. The varied mechanisms that may be responsible for fracture (e.g. fatigue, creep, overload) will result in cracks that display ductile and brittle elements, often in patterns unique to the relevant mechanism. These phenomena will be discussed in detail in Section 4.3.

Ductile fracture generally occurs by slow crack propagation with large observable plastic deformation prior to fracture. In metals, ductile fracture most often has a gray, fibrous and dimpled cup and cone appearance. The phenomenon of ductile fracture may be divided into the following stages:

- Initial necking (region of plastic deformation);
- Small cavities (micro-voids) form in the necked region;
- Cavities grow and coalescence to form a crack;
- Fast crack propagation;
- Final shear fracture (cup and cone).

Recall that dislocations (due to shear loads) are the primary mechanisms responsible for plastic deformation. Ductile materials therefore tend to fail along planes oriented parallel to the maximum shear stress direction.

Brittle fracture occurs at or below the elastic limit of a material. Small, localized areas of plasticity may however exist due to very high stress concentrations near the crack tip. Brittle fracture is characterized by rapid crack propagation without significant plastic deformation. Since plastic deformation is insignificant, the energy required for brittle crack propagation is substantially less than that required for ductile crack propagation. The rate of crack propagation for brittle fracture will increase with increasing strain rate, decreasing temperature, or by increasing the density of local stress concentrations (notches, scratches). Brittle fracture may be transgranular or intergranular; this depends upon whether the grain boundaries are stronger or weaker than the grains. In metals, brittle fracture occurs along crystallographic planes and across the grain boundaries (transgranular); this is known as "cleavage fracture." Figure 1.8 illustrates the difference in appearance of ductile and brittle fractures. The samples can be assumed to be subjected to tensile loading. Note that the ductile failure angle is $\approx 45^0$ as this is a plane of maximum shear relative to the principal plane. The sample is subjected to tensile loading only, so the principal

Ductile fracture

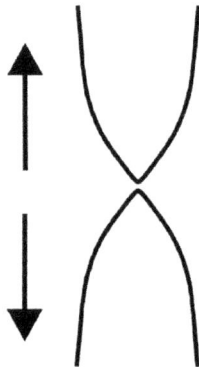

Brittle fracture
Shows little plastic deformation

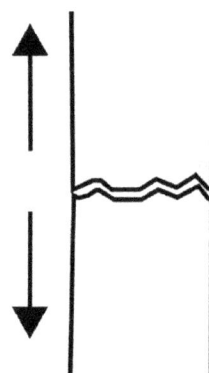

FIGURE 1.8   Ductile vs. brittle fracture.

planes are parallel and perpendicular to the tensile load. Brittle fractures (cleavage) are more dangerous than ductile fractures as they occur suddenly and rapidly, without warning or prior plastic deformation.

1.11   FRACTURE OF NONMETALS

Nonmetals include ceramics, polymers, and composites. Ceramics are inorganic compounds comprised of a metallic element and a non-metallic element. The interatomic bond is typically an ionic one; however, some ceramics have bonds that are mostly ionic with some covalent character. Most ceramics are crystalline in nature; examples include clay, porcelain, cement, metal oxides, nitrides, and carbides. Various types of glasses are non-crystalline ceramics. Ceramic material is more brittle than metal and is therefore subject to brittle fracture. Fracture of crystalline ceramics is usually transgranular (see Figure 1.4) and may most likely occur along high-density crystallographic planes (cleavage planes). The resulting fracture surface will have a mirror-like region surrounding the origin of the crack, bordered by a misty region containing numerous micro-cracks. Ceramics are strong in compression and weak in shear and tension; they are chemically inert and do not conduct heat or electricity. The fracture toughness of ceramics is measured using the stress intensity critical value, K_C, derived from LEFM. However, the dispersion of fracture property values for a single class of ceramics is significant due to the presence of multiple microscopic flaws (micro-cracks, internal pores, and atmospheric contaminants). The occurrences of these flaws cannot currently be controlled very well as the ceramic cools during the manufacturing process. The variability in the fracture toughness of ceramics lends the fracture analysis to a statistical approach. Generally statistical predictions coupled with experimental data are used to determine the risk of fracture for ceramic materials.

Polymers are chemicals that are comprised of many repeating units. Each unit typically consists of carbon, hydrogen, and sometimes one or more of the following elements: phosphorus, oxygen, silicon, nitrogen, chlorine, sulfur, and/or fluorine. The chain of units may be 1D, 2D, or 3D. Polymeric materials are generally significantly more ductile than metals. These materials therefore undergo ductile fracture; however, numerous micro-voids may form prior to necking. A crack is initialized through the coalescence of voids, it then propagates by the growth of voids ahead of the advancing crack tip. As many small voids are produced in the material, it typically shows a cloudy appearance prior to necking. While increased ductility in metals implies a greater fracture toughness, relatively ductile polymers can have lower absolute toughness values than some metals that fail with minimal plasticity. *We therefore cannot always assume that high ductility implies high fracture toughness.*

Composites are comprised of two or more materials which remain chemically and physically distinct. Therefore, the materials do not dissolve or chemically combine with one another. Natural composites include bone, wood, and straw. Often the properties of each constituent are quite dissimilar, therefore the composite will exhibit properties which are a combination of the constituent properties. Typically, one material surrounds and binds fibers or particulates of the other. In the fabrication of conventional laminate composites, layers of oriented or randomly oriented fibers (2D), bound by some matrix, are stacked on each other, and then subjected to high pressure-temperature cycles. The binding material or matrix is typically a polymer such as epoxy resin. Some examples of laminate composites include carbon fiber reinforced composites (CFRMs), ceramic matrix composites (CMMs), and metal matrix composites (MMCs). The desirability of composites lies in the fact that their mechanical properties can be tailored for specific applications by altering fiber orientation, thickness, size, the number of layers used, and/or the constituent materials. Engineered composites have therefore been used successfully in a wide variety of applications ranging from sporting equipment to primary aircraft structures.

Laminated composites can however fail structurally. The most common failure mechanism is the separation of the laminates; this is referred to as delamination. Other failure processes may involve matrix cracking, void growth, matrix/fiber interface debonding, intra- and/or interlaminar damage in the form of transverse cracking, and fiber splitting [9, 10]. From a fracture mechanics perspective, the fracture toughness of both the fiber and the matrix need to be considered in any analysis. These will both vary with fiber orientation for a single lamina or ply.

1.12   SOFTWARE USE IN FRACTURE MECHANICS

Many computer-aided engineering (CAE) software packages now include fracture mechanics add-ons, for linear, elastic-plastic, as well as plastic fracture mechanics. Additionally, there are stand-alone software packages that allow CAE data to be imported and organized for input to a fracture-based analysis. Typical inputs/outputs will be discussed in more detail in subsequent chapters. For direct crack growth, some growth criteria will first be specified; the software will then predict whether the crack -initiates, whether it grows, and how far the crack path is. In low cycle fatigue, items such as crack path and shape may be determined based on the prescribed load cycle and growth increment. In high cycle fatigue, some representative loading will first be prescribed. The software may then determine the crack path and shape, and the number of fatigue cycles as a function of crack length.

NASGRO [11] was originally developed by NASA in the early 1980s to: perform fracture control for manned space programs; calculate stress intensity factors, crack growth life, and critical crack size; as well as curve-fit fatigue crack growth data. AFGROW [12], originally called MODGRO, is also a crack growth life prediction program. MODGRO was originally developed by Ed Davidson of the Aeronautics Systems Division (ASD) at Wright Patterson Air Force base in the early 1980s. AFGROW is widely used by the US Air Force and within the aerospace industry in general. Other popular software packages at the time of writing include ABAQUS [13], ANSYS [14], MechaniCalc. Inc [15], BEASY [16], WARP3D (open source) [17], CrackWISE® fitness-for-service (FFS) [18], ZENCRACK [19], and ESACRACK [20]. Figure 1.9 shows an example for selecting model geometry within the AFGROW crack growth software package. The inputs will be discussed in subsequent chapters.

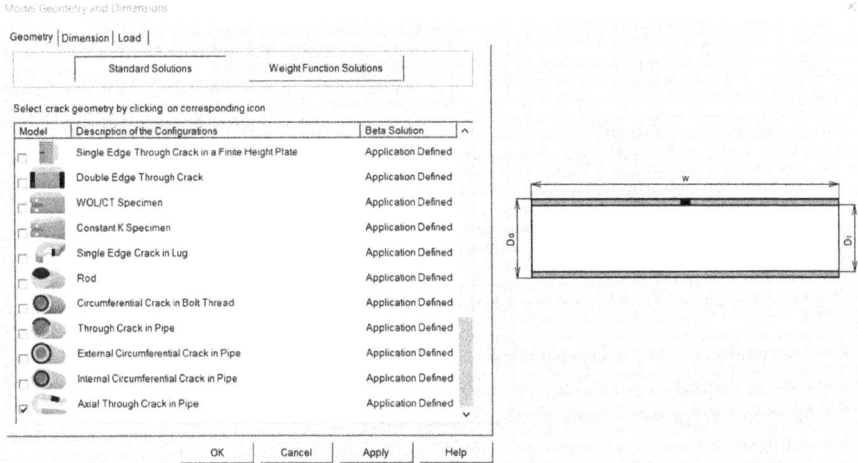

FIGURE 1.9 Model geometry options in AFGROW 5.3 crack growth software package.

REVIEW AND PRACTICE PROBLEMS

1.1. By drawing a 2-D quadrilateral element in the x-y plane, show that the enforcement of moment equilibrium results in $\sigma_{xy} = \sigma_{yx}$.

1.2. Explain what is meant by the term fracture toughness? How would you expect fracture toughness to be influenced by temperature for the classes of materials below? Why?
   a. Metals
   b. Polymers
   c. Ceramics

1.3. List which fabrication methods might promote anisotropy for metals and polymers. Explain why anisotropy results from these fabrication processes.

1.4. Several aviation examples of fracture failure were presented in this chapter. Cite three recent fracture failures from each of the categories below. Were these due to fatigue, environmentally assisted fatigue, fatigue with other micro-imperfections, static, or dynamic overload?
   a. Rail
   b. Large pressure vessels (nuclear, oil, food)
   c. Pipelines (natural gas, oil)
   d. Civil infrastructure (bridges, buildings, etc.)

1.5. The ability to determine what scenarios can be approximated as plane stress or plane strain is essential for fracture mechanics. Indicate which situation below is plane stress or plane strain and provide justification.
   a. A pressurized aircraft fuselage
   b. A dike wall

  c. Side walls in an open channel
  d. Surface of an automobile drive shaft
  e. Tunnels subjected to uniform transverse loads

Elementary Mechanics of Materials problems are provided here for review purposes. A strong base in the understanding of Mechanics of Materials concepts is recommended before attempting to study Fracture Mechanics. A brief review of the key concepts in Mechanics of Materials is presented in Appendix A.

ELEMENTARY MECHANICS OF MATERIALS PROBLEMS

1.6. A cylindrical beam is subjected to an axial compressive force of 55,000 N. The maximum allowable axial stress in the beam is 25 MPa. What is the diameter required to ensure that the beam stays under the maximum allowable stress?

1.7. An elastic beam is 20 m long and has a 10 cm diameter. An axial force of 1 MN is applied to the end. Once the force is applied, the beam is measured and found to be 20.012 m long. Assuming the beam has not exceeded its elastic limit, what is the modulus of elasticity for this beam?

1.8. The part shown in Figure 1.10 is subjected to two forces as shown below. There is a pin connection at A. What is the shear stress experienced by the pin at A?

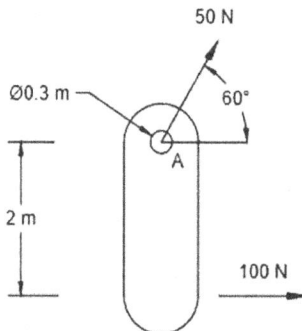

FIGURE 1.10 Pin-connected plate subjected to two concentrated loads.

1.9. A two-part square beam has a force applied to it as shown in Figure 1.11. The modulus of elasticity of the thin portion is 30,700 ksi, and the modulus of elasticity of the thick portion is 25,400 ksi. What is the total deformation of the beam?

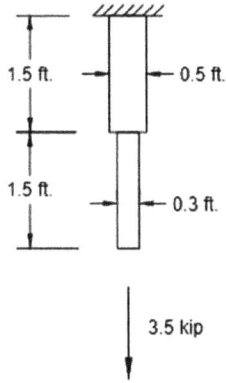

FIGURE 1.11   A two-part square beam with an applied concentrated load.

1.10. A solid steel shaft of 30 mm diameter has a maximum allowing shear stress
of 70 MPa. What is the maximum torque the bar can resist?

1.11. Figure 1.12 shows a beam's cross-section. Find the location of the centroid
and the moment of inertia about its neutral axis. Determine the maximum
compressive and tensile stresses in the beam when it is subjected to a bending moment of 25,000 Nm.

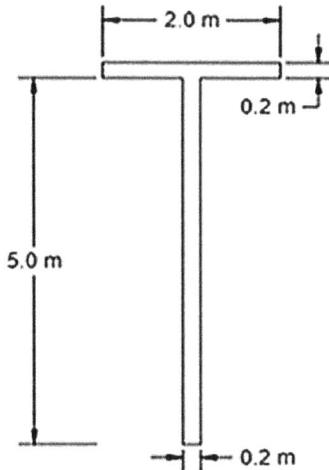

FIGURE 1.12   T-beam cross-section.

1.12. When a cylindrical bar is subjected to torsion as shown in Figure 1.13, the
circumferential stress element is in a state of pure shear. This means the

element is subjected to shear stresses only. Use Eqns (A.3, A.5) in Appendix A to determine the principal stresses $(\sigma_{x'x'}, \sigma_{y'y'})$ in terms of the shear stress σ_{xy}, and the principal angle, θ_p.

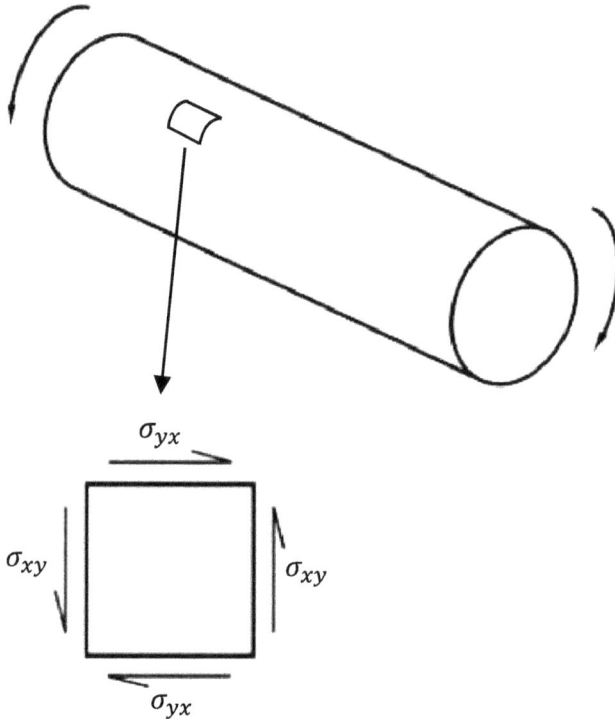

FIGURE 1.13 Cylindrical bar under torsional load (pure shear).

REFERENCES

[1] U. States, "Final Report of a Board of Investigation to Inquire into the Design and Methods of Construction of Welded Steel Merchant Vessels," US Government Printing Office, Washington DC, 1947.

[2] P. Thompson, "How Much Did the Liberty Shipbuilders Learn? New Evidence for an Old Case Study," *Journal of Political Economy*, vol. 109, no. 1, pp. 103–137, 2001.

[3] K. Zappas, "Constance Tipper Cracks the Case of the Liberty Ships," *JOM*, vol. 67, no. 12, pp. 2774–2776, 2015.

[4] "The President, 1958–9—Sir Arnold Hall, M.A., F.R.S., F.R.Ae.S.," *The Aeronautical Journal*, vol. 62, no. 569, pp. xviii–xix, 1958.

[5] C. Tipper, *The Brittle Fracture Story*, Cambridge: Cambridge Unversity Press, 1962.

[6] "NTSB/AAR-89/03," National Transportation and Safety Board, Washington DC, 1989.

[7] NTSB/AAR-18/01, "Uncontained Engine Failure and Subsequent Fire American Airlines Flight 383 Boeing 767-323, N345AN," National Transportation and Safety Board, Washington DC, 2018.

[8] National Transportation and Safety Board, "NTSB Aviation Accident Final Report Accident #DCA17MA022," NTSB, Washington DC, 2016.

[9] M. Kashtalyan and C. Soutis, "Analysis of composite laminates with intraand interlaminar damage.," *Progress in Aerospace Sciences*, vol. 41, no. 2, pp. 152–173, 2005.

[10] I. Hanhan, A. M. Ortiz-Morales, J. J. Solano and M. D. Sangid, "Slow crack growth in laminate composites via in-situ X-ray tomography and simulations," *International Journal of Fatigue*, vol. 155, no. 106612, 2022.

[11] Southwest Research Institute, NASA, NASGRO Consortium, NASGRO version 9.2, San Antonio, TX: Southwest Research Insitute, 2020.

[12] US Air Force, Lextech Inc., AFGROW 5.03.05.24, LexTech, Inc, Centerville, OH, 2020.

[13] Dassault Systemes, "ABAQUS/3DS," Dassault Systemes Software Corporation, Velizy-Villacoublay, 2020.

[14] Anysys Simulation Software, "Ansys," Ansys Simulation Software Company, Canonsburg, 2020.

[15] MechaniCalc, "Fracture Mechanics Calculator," MechaniCalc, Inc, https://mechanicalc.com/reference/fracture-mechanics, 2014–2022.

[16] BEASY USA-Computational Mechanics International Inc., "Fracture and Crack Growth Simulation," BEASY Software and Services, Billerica, 2022.

[17] Open Source, "Warp 3D-Open Source Code for the Aanalysis of 3D Solids," 18.3.1 Current Release: Build 4210, http://www.warp3d.net/, 2021.

[18] TWI, "Fitness for Service Assessment with Crack Wise," Cambridge: TWI, 2022.

[19] Zentech International Limited, Zencrack, 590B Finchley Road, London NW11 7RX.: Zentech International Limited, 2021.

[20] European Space Agency (ESA), "ESACRACK Version 4.3.1a," ESA, Paris, 2012.

2 Fundamentals of Linear Elastic Fracture Mechanics
Basic

OBJECTIVES

After studying the first part of the chapter (Sections 2.1–2.6), the student should be able to:

1. Understand and apply the basic fracture stress model that is based on inter-atomic forces.
2. Describe the stress intensity factor and understand how it differs from a stress concentration factor.
3. Know and be able to describe how flaw size and finite length affects stress intensity.
4. Understand how to apply superposition and multiplicity in Linear Elastic Fracture Mechanics (LEFM) problems.

2.1 EARLY THEORETICAL FOUNDATIONS

There are numerous resources published as print and online that provide detailed derivations of the governing LEFM equations. As we are primarily concerned with application, we will derive only a few classical equations; however, most governing equations will be stated without proof. It is still though extremely important to understand the assumptions behind each equation. These will be reinforced throughout each discussion and revisited during problem solving.

Figure 2.1 shows the change in interatomic force with atomic spacing. The interatomic force–displacement relationship beyond the equilibrium spacing can be approximated as one-half period of a sine wave (wavelength = 2λ) and described by the equation:

$$F = F_{max} \sin \frac{\pi x}{\lambda}$$

where F represents the interatomic force between two atoms, F_{max} is the maximum value of the interatomic attractive force as the atoms move further apart, x is the interatomic spacing, and λ is half the wavelength of the sine curve.

DOI: 10.1201/9781003052050-2

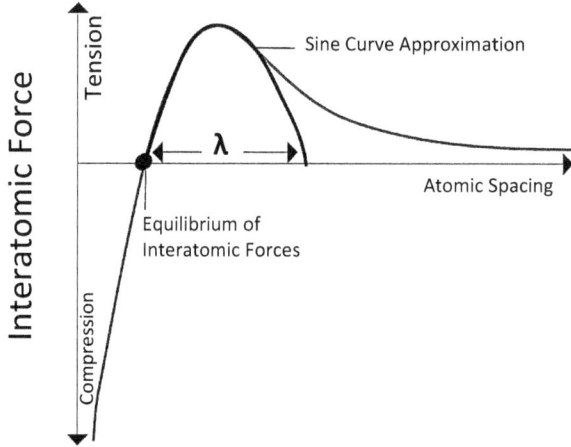

FIGURE 2.1   Interatomic force vs. atomic spacing.

Assuming small displacements, $\sin\dfrac{\pi x}{\lambda} \cong \dfrac{\pi x}{\lambda}$, so we can write

$$F = F_{max}\frac{\pi x}{\lambda} = \left(\frac{F_{max}\pi}{\lambda}\right)x$$

A conversion from force to stress can be obtained by dividing both sides by the cross-sectional area of the bond:

$$\frac{F}{A} = \left(\frac{F_{max}\pi}{\lambda A}\right)x$$

$$\sigma = \left(\frac{\sigma_{max}\pi}{\lambda}\right)x$$

A conversion of displacement to strain can be obtained by dividing the displacement by the equilibrium spacing x_0.

Therefore

$$\sigma = \left(\frac{\sigma_{max}\pi x_0}{\lambda}\right)\frac{x}{x_0}$$

$$\Rightarrow \sigma = \left(\frac{\sigma_{max}\pi x_0}{\lambda}\right)\varepsilon$$

where ε represents linear strain. The elastic modulus, E, is therefore given by

$$E = \frac{\sigma_{max}\pi x_0}{\lambda}$$

$$\Rightarrow \sigma_{max} = \frac{E\lambda}{\pi x_0} \tag{2.1}$$

σ_{max} represents the cohesive stress of the material (i.e. stress due to interatomic forces).

Experimentally obtained values for the cohesive stress were found to be three to four orders of magnitude less than that predicted by Eqn. (2.1). This led to the postulation that the presence of flaws magnified the local stress, in other words the flaws produced a local "stress concentration" effect.

2.2 STRESS CONCENTRATION

Consider a plate subjected to stress as shown in Figure 2.2. Holes within any geometry result in a diminished area of material available for the cross-section. The lines of stress are therefore more concentrated within the smaller cross-section. The local stress will be some multiple of the remote stress. This multiplying factor is called the stress concentration factor, which is dependent on geometry only. Stress concentration occurs for any geometrical reduction of cross-sectional area; therefore notches, grooves, and fillets all result in stress concentrations as well.

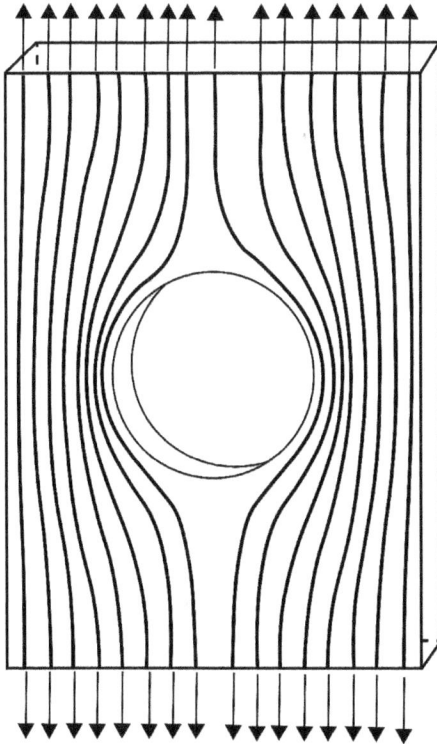

FIGURE 2.2   Stress concentration.

The earliest work in fracture mechanics that developed this idea of stress concentration was performed by Kirsch [1], Inglis [2], and Griffith [3]. Kirsch demonstrated that local stresses in the vicinity of a circular hole in a thin infinite plate subjected to tensile loads can significantly exceed the remotely applied stress. The term "infinite" requires that the plate width and height $\gg 2r$, where r is the radius of the hole. Therefore, the hole is not influenced by the plate width or height. Figure 2.3 shows the tensile stress distribution along the transverse axis of symmetry through the hole in a thin plate subjected to tension. As we move towards a distance equal to one hole diameter away from the edge of the hole, the tensile stress decreases from $3\sigma_\infty$ to σ_∞.

Inglis conducted quantitative work examining an elliptical hole in an infinite plate subjected to tension. This work resulted in the development of two important algebraic relationships that provided insight regarding the influence of geometric properties on the local stress distribution. Inglis showed, using linear elastic theory, that the stress, σ, in the vicinity of the elliptical hole as shown in Figure (2.3b) is given by

$$\sigma = \sigma_\infty \left(1 + 2\frac{a}{b}\right) \tag{2.2}$$

where a and b are one-half of the major and minor axes lengths, respectively. The stress concentration factor in this case is therefore $1 + 2\left(\dfrac{a}{b}\right)$. The radius of curvature, ρ, at point A of the ellipse is given by

$$\rho = \frac{b^2}{a}$$

$$\Rightarrow b = \sqrt{\rho a}$$

Substituting this expression for b in Eqn. (2.1), we obtain

$$\sigma = \sigma_\infty \left(1 + 2\sqrt{a/\rho}\right)$$

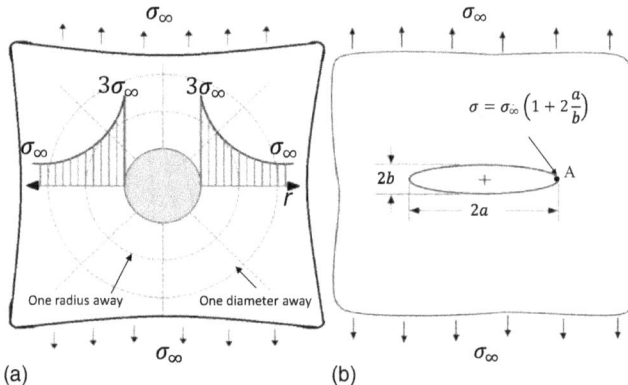

FIGURE 2.3   Thin plate with a circular hole (a) and an elliptical hole (b), subjected to uniaxial stress.

If the ellipse is viewed as a sharp crack, then $a \gg b$ therefore

$$\frac{a}{b} \gg 1 \Rightarrow \frac{2a}{b} \gg 1 \Rightarrow 2\sqrt{\frac{a}{\rho}} \gg 1$$

Equation (2.2) applied to a sharp crack (i.e. $a \gg b$) can now be written as

$$\sigma = 2\sigma_\infty \sqrt{\frac{a}{\rho}} \tag{2.3}$$

Equation (2.3) leads to the conclusion that the stress at the crack tip increases as the crack becomes sharper. In fact, the equation implies that the stress near the crack tip will approach infinity as the radius of curvature, ρ, approaches zero. This inverse relationship is observed physically; however, the stress does not become infinite for sharp cracks. Recall that the derivation of Eqn. (2.3) is based on linear elastic theory; once the stress exceeds the material yield strength, localized plastic deformation occurs at the crack tip thereby invalidating linear elastic predictions for this region. The plastic deformation also leads to blunting at the crack tip.

Equation (2.3) was an important derivation that offered insight regarding the sharpness of the crack tip and the intensity of the stress magnification in its vicinity. Other researchers at the time adopted a strength of materials approach at the atomic level and developed another important relationship that offers insight into the relationship between fracture stress, modulus of elasticity, and surface energy. This derivation is described below.

Recall that the interatomic potential energy at the equilibrium distance x_0 is determined by integrating the atomic force over the distance from x_0 to ∞, that is

$$\int_{x_0}^{\infty} F \, dx$$

The energy per unit area required to produce fracture (break an atomic bond) is therefore

$$\int_{0}^{\lambda} \sigma_{max} \sin \frac{\pi x}{\lambda} \, dx = \frac{2\sigma_{max}\lambda}{\pi}$$

The surface energy, γ_s, is the energy per unit area needed to create a new surface. Since two new surfaces are created during fracture, γ_s is one-half of the total fracture energy per unit area; therefore

$$\gamma_s = \frac{\sigma_{max}\lambda}{\pi} \tag{2.4}$$

Eliminating λ between Eqns (2.1) and (2.4), we obtain

$$\sigma_{max} = \sqrt{\frac{E\gamma_s}{x_0}} \qquad (2.5)$$

If we assume that the minimum radius at the crack tip is the equilibrium spacing, that is $\rho = x_0$, and that fracture occurs when the stress at the crack tip reaches some critical value, that is $\sigma = \sigma_{max}$, then Eqn. (2.3) can be combined with Eqn. (2.5) to derive an expression for the remote stress at failure, σ_f:

$$\sigma_f = \sqrt{\frac{E\gamma_s}{4a}} \qquad (2.6)$$

Equation (2.6) assumes the material is a continuum and uses a strength of materials approach. The material may not necessarily be a continuum as other imperfections may exist near the crack. Additionally, it defines the exceedance of some maximum stress value at the crack tip only. This single high stress point may not be enough to cause failure. If the material is subjected to a high stress gradient it may not necessarily fracture due to a single high stress point.

Example 2.1

What is the maximum stress occurring at the tip of an elliptical internal crack with a length of 0.026 mm and a radius of curvature of 3.5×10^{-4} mm? The material is subjected to a stress of 300 MPa.

The stress occurring at the tip of the elliptical crack is defined by:

$$\sigma = 2\sigma_\infty \sqrt{\frac{a}{\rho}}$$

$$\sigma = 2(300\,\text{MPa})\left(\frac{\frac{0.026\,\text{mm}}{2}}{(3.5 \cdot 10^{-4}\,\text{mm})}\right)^{1/2} = 3656.7\,\text{MPa}$$

Note: the length of the crack was halved because a is defined as half the length of the crack length for the case of a crack in an infinite plate.

Important: We obtained our answer by directly applying Eqn. (2.3). Is this really applicable to this problem? What questions should we have asked before applying this equation?

1. "The material is subjected to a stress of 300 MPa." How is this stress being applied? Equation (2.3) was derived assuming tensile loading for a thin plate that can be modeled as infinite. So how do we know the stress being applied is tensile?
2. Was the load applied uniformly? Can the plate be considered infinite?
3. Is $a \ll b$? How do we know the plate is thin?
4. Examine Figure 2.3. Is the elliptical crack for this problem close to an edge? If yes, we could not apply Eqn. (2.3).

Concept Challenge 2.1

In cases where cracks are found in service and repairs cannot be done immediately, one engineering solution to reducing the crack growth rates is to drill
holes at the crack tips. Use Eqn. (2.3) to explain why this works at least as a
temporary solution.

2.3 PLANE STRAIN OR PLANE STRESS PROBLEMS

There can be no stress on a free surface. Therefore, the through-the-thickness stress,
σ_{zz}, must be zero for the surfaces perpendicular to both crack planes. However, σ_z will
be non-zero at the mid-thickness plane of the crack tip. For a thin plate, as shown in
Figure 2.4(a), the increase in σ_{zz} from the free surfaces to the mid-thickness plane will
be small, so we can approximate a plane stress condition:

$$\sigma_{zz} \approx 0 \tag{2.7}$$

If the plate is thick, as shown in Figure 2.4(b), σ_{zz} at the mid-thickness plane can
develop a larger value, creating a tri-axial tensile stress state at the crack tip. This
restricts straining in the z-direction; therefore, we can approximate a plane strain
condition:

$$\sigma_{zz} \approx v\left(\sigma_{xx} + \sigma_{yy}\right) \tag{2.8}$$

Concept Challenge 2.2

What does the stress tensor, Eqn. (1.1), look like for the plane stress condition?
What does it look like for the plane strain condition?

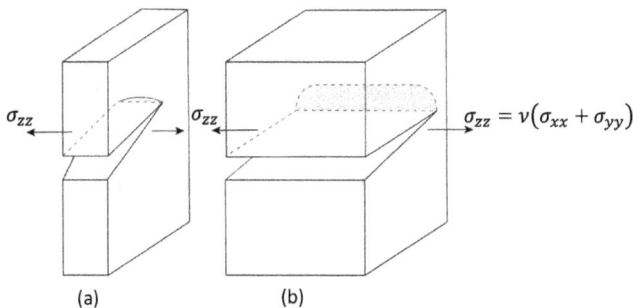

FIGURE 2.4 Through the thickness stress for (a) plane stress and (b) plane strain.

2.4   STRESS INTENSITY FACTOR

The stress intensity factor, K, was introduced in Section 1.8 and will be now fully defined in this section. An analytical expression that relates K with the state of stress at the crack tip was first developed in 1957 by George R. Irwin [4]. The state of stress in *any* linear elastic cracked body was found to have the form:

$$\sigma_{ij} = \left(\frac{K}{\sqrt{2\pi r}} \right) f_{ij}(\theta) + H.O.T. \tag{2.9}$$

where θ is the angle of inclination of the stress element from the crack axis and r is the radial distance from the crack tip, as shown in Figure 2.5; f_{ij} is a dimensionless function of θ; and *H. O. T.* means higher order terms. From Eqn. (2.9), the units of K must be *stress* $\cdot \sqrt{length}$; in the metric system this is typically MPa$\sqrt{m}$. The (*H.O.T.*) tend to zero, or are finite as $r \to 0$. Note that the lead term is proportional to $1/\sqrt{r}$, therefore Eqn. (2.9) predicts that a *stress singularity* will exist at the crack tip.

Crack loading is categorized as having three modes: Mode I opening, Mode II in plane shear, and Mode III out of plane shear. These modes are shown in Figure 2.6. Each mode of loading will result in a unique value for K, associated with that mode only. The K values for each mode are represented as K_I, K_{II}, and K_{III}. Equation (2.9) will therefore result in different stress states for different modes. Table 2.1 provides expressions for the stress ahead of the crack tip for Modes I and II. If the crack is under *mixed mode loading*, the stress states due to each mode may be calculated and superposed to obtain the resulting stress.

2.4.1   *K* AND GLOBAL BEHAVIOR

Since K applies to linear elastic material where all stress components in all locations are proportional to the applied loading, K is also proportional to the applied stresses.

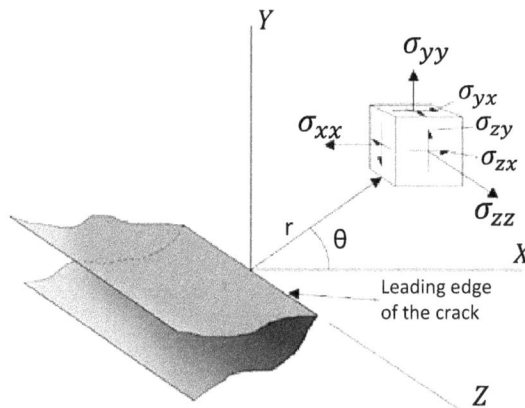

FIGURE 2.5   Stress element near crack tip.

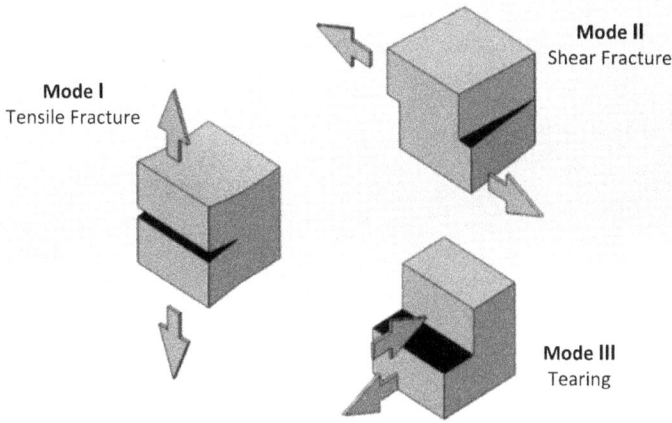

FIGURE 2.6 Crack loading modes.

TABLE 2.1
Stress Ahead of Crack Tip for Modes I and II

	Mode I	Mode II
σ_{xx}	$\dfrac{K_I}{\sqrt{2\pi r}}\cos\left(\dfrac{\theta}{2}\right)\left[1-\sin\left(\dfrac{\theta}{2}\right)\sin\left(\dfrac{3\theta}{2}\right)\right]$	$-\dfrac{K_{II}}{\sqrt{2\pi r}}\sin\left(\dfrac{\theta}{2}\right)\left[2+\cos\left(\dfrac{\theta}{2}\right)\cos\left(\dfrac{3\theta}{2}\right)\right]$
σ_{yy}	$\dfrac{K_I}{\sqrt{2\pi r}}\cos\left(\dfrac{\theta}{2}\right)\left[1+\sin\left(\dfrac{\theta}{2}\right)\sin\left(\dfrac{3\theta}{2}\right)\right]$	$\dfrac{K_{II}}{\sqrt{2\pi r}}\sin\left(\dfrac{\theta}{2}\right)\cos\left(\dfrac{\theta}{2}\right)\cos\left(\dfrac{3\theta}{2}\right)$
σ_{xy}	$\dfrac{K_I}{\sqrt{2\pi r}}\cos\left(\dfrac{\theta}{2}\right)\sin\left(\dfrac{\theta}{2}\right)\cos\left(\dfrac{3\theta}{2}\right)$	$\dfrac{K_{II}}{\sqrt{2\pi r}}\cos\left(\dfrac{\theta}{2}\right)\left[1-\sin\left(\dfrac{\theta}{2}\right)\sin\left(\dfrac{3\theta}{2}\right)\right]$
σ_{zz}	0 (Plane Stress) v $(\sigma_{xx}+\sigma_{yy})$ (Plane Strain)	0 (Plane Stress) v $(\sigma_{xx}+\sigma_{yy})$ (Plane Strain)
σ_{xz}, σ_{yz}	0	0

Recall K also depends on $\sqrt{length}$; the relevant length in fracture mechanics would be the half-length of the crack, a. For Mode I loading, $K = K_I$. Closed form solutions and experimental and numerical techniques to determine K_I yield:

$$K_I = Y\sigma\sqrt{\pi a} \qquad (2.10)$$

where Y is a parameter that depends on geometry and loading for an infinite or semi-infinite plate. For a finite plate (i.e. a is NOT $\ll W$), Y will therefore also depend on the ratio a/W. In the case of a central through crack in an infinite plate subjected to tensile loading, $Y = 1$. Table 2.2 provides a few selected expressions for K_I for specific

TABLE 2.2
Selected Stress Intensity Factor (SIF) Simple Cases

Case	Diagram	SIF Expression
Through crack in infinite plate subjected to tensile loading		$K_I = \sigma \sqrt{\pi a}$
Through crack (normal to crack is oriented at an angle β to stress axis)	β	$K_I = \sigma \cos^2 \beta \sqrt{\pi a}$ $K_{II} = \sigma \sin \beta \cos \beta \sqrt{\pi a}$
Edge crack in a semi-infinite plate subjected to tensile loading		$K_I = 1.12 \, \sigma \sqrt{\pi a}$
Penny shaped crack in an infinite medium		$K_I = \dfrac{2}{\pi} \sigma \sqrt{\pi a}$

geometries and loading. Note the Y value of 1.12 for the through crack at an edge can be regarded as a surface flaw correction factor. The free surfaces above and below the crack lower the resistance to crack opening, thereby increasing K_I. This correction appears in expressions for both through and surface cracks when they are at an edge.

2.4.2 FLAW SHAPE

The shape of the flaw for surface or embedded cracks also influences the stress intensity value, K_I. In these cases, a flaw shape factor, Q, is incorporated into the SIF equation. For example, the SIF for a semi-elliptical surface crack in an infinite plate is given by

$$K_1 = 1.12\sigma \sqrt{\pi \frac{a}{Q}} \tag{2.11}$$

where $Q = f\left(\dfrac{a}{2c}, \sigma\right)$, 2c *is the length of the elliptical crack's major axis and a is* the half-length of the minor axis. Since Q is a function of the applied stress and aspect ratio, it is typically plotted vs. $\dfrac{a}{2c}$ at various stress levels for different materials. Figure 2.7 shows a typical plot of $\dfrac{a}{2c}$ vs. Q used to determine Q.

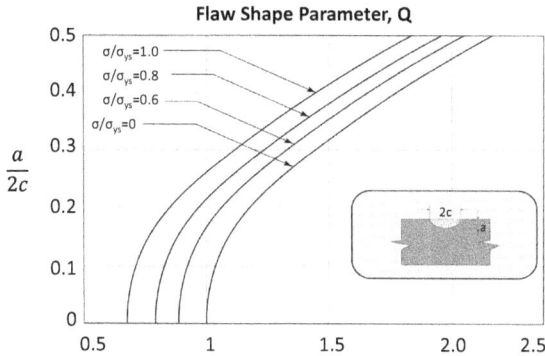

FIGURE 2.7   Typical crack aspect ratio vs. flaw shape parameter.

Concept Challenge 2.3

Suppose an infinite thin plate with a central crack is loaded biaxially. Would you expect the fracture toughness to change compared to an infinite thin plate with a central crack loaded axially (perpendicular to the crack direction)? Why?

2.4.3   MULTIPLICITY OF GEOMETRY FACTOR, Y

A flaw may be oriented such that more than one known geometry correction factor is applicable. In this case, the absolute geometric factor can be approximated by taking the product of the individual geometric factors. Therefore, if geometry factors Y_1, Y_2, Y_3 are applicable to a single crack, then the stress intensity factor K is approximated as:

$$K \approx Y_1 \cdot Y_2 \cdot Y_3 \cdot \sigma \sqrt{\pi a} \tag{2.12}$$

For example, the circular corner crack shown in Figure 2.8 is a combination of two surfaces (surface flaw correction, $Y = 1.12$) and an embedded circular flaw ($Y = 2/\pi$) (see Table 2.2). The stress intensity may therefore be approximated by

$$K \approx (1.12)^2 \left(\frac{2}{\pi} \right) \sigma \sqrt{\pi a} \tag{2.13}$$

2.4.4   STRESS INTENSITY VS. STRESS CONCENTRATION FACTOR

The *stress concentration factor* is the ratio of far field or applied gross stress to local stress. It is a dimensionless quantity that depends on the flow of material within a given geometry. Recall the stress concentration factor tends to infinity as the radius

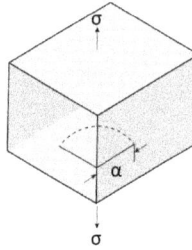

FIGURE 2.8   Circular corner crack.

TABLE 2.3
Stress Concentration vs. Stress Intensity

Stress Concentration Factor	Stress Intensity Factor
Holes, fillets, changes in cross sectional geometry	Cracks only
Geometry dependent	Geometry and load dependent
No units	Has units, typically MPa$\sqrt{m}$ or Ksi$\sqrt{in}$
Dependent on orientation of geometric feature e.g. elliptical hole in a tensile loaded plate	Dependent on crack orientation
Not dependent on geometric feature size, e.g. the stress concentration for a hole in an infinite plate is 3 regardless of hole size	Dependent on crack size
Does not define state of stress at a point	Completely defines state of stress at crack tip

of curvature of the geometrical feature decreases. The *stress intensity factor* has the following features:

- It is applicable to cracks only;
- It depends on geometry, crack length, and crack orientation;
- It has a strong dependence on loading;
- It is a function of the radius at the crack tip;
- It is *not* the ratio of far field stress to the local stress;
- It can completely describe the state of stress at the crack tip and has units.

The differences between the stress concentration factor and the stress intensity factor are shown in Table 2.3. It is important to note that *the stress concentration factor may apply to a crack as an additional geometry factor (Y) if the crack is in the vicinity of the relevant geometric feature.*

2.5   FINITE SIZE CORRECTION

Thus far, we have explored semi-infinite and infinite plates. What happens in the case of finite plates, that is the crack size is not small compared to the plate geometry? In this case, the crack tip conditions will be influenced by the external boundaries.

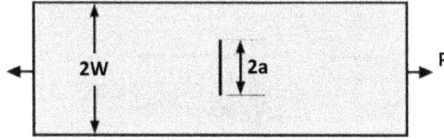

FIGURE 2.9 Center-cracked tension (CCT) specimen.

One can visualize the lines of force shown in Figure 2.2 to align with the side boundary earlier, the closer the boundary is to the crack. This greater compression of the lines of force results in higher stress values. Solutions for the finite width correction have been obtained through polynomial fits to computer-based solutions. For example, the center-cracked tension (CCT) specimen shown in Figure 2.9 has the solution

$$K_I = \sqrt{\sec\left(\frac{\pi a}{2W}\right)}\left[1 - 0.025\left(\frac{a}{W}\right)^2 + 0.06\left(\frac{a}{W}\right)^4\right]\sigma\sqrt{\pi a} \qquad (2.14)$$

where W is half the plate width. Note the term preceding $\sigma\sqrt{\pi a}$ is the geometry correction factor Y.

Common formulae that include finite width corrections for selected load/geometry conditions are to be found in Appendix B. Solutions for common test specimens are provided in Appendix C.

Example 2.2

A large sheet containing a 30 mm long central crack fractures when subjected to an applied tensile stress of 450 MPa. Determine the fracture load of a similar sheet with a 60 mm crack.

SOLUTION

That the sheet is "large" indicates that we can confidently apply the solution for a semi-infinite plate, that is the width is large compared to the crack size, so finite width effects may be ignored. We also know that $Y = 1$ since it is a through crack and loading is tensile, and we know it is thin as the word "sheet" implies this. The governing equation is

$$\sigma_f = \frac{K_{Ic}}{\sqrt{\pi a}}$$

If K_{Ic} is constant, then

$$\sigma_f \propto \frac{1}{\sqrt{a}}$$

$$\frac{\sigma_{f1}}{\sigma_{f2}} = \frac{\sqrt{a_2}}{\sqrt{a_1}}$$

Therefore

$$\sigma_{f2} = \sigma_{f1}\sqrt{\left(\frac{a_1}{a_2}\right)} = 450\sqrt{\left(\frac{15 \times 10^{-3}}{30 \times 10^{-3}}\right)} = 318.2 \text{ MPa}$$

Alternatively, we can first calculate K_{Ic}

$$K_{Ic} = 450\sqrt{\pi\left(15 \times 10^{-3}\right)} = 97.7 \text{ MPa}\sqrt{m}$$

Then determine σ_f for a 60 mm crack, $\sigma_f = \dfrac{K_{Ic}}{\sqrt{\pi\left(30 \times 10^{-3}\right)}} = \dfrac{97.7}{0.307} = 318.2 \text{ MPa}$

For discussion: Which method will yield a more accurate answer in general? Why?

2.6 SUPERPOSITION

The "linear" description within the term "linear elastic materials" implies that the individual solutions for stress, strain, or displacement due to individual loading may be added to obtain solutions when these loads are combined, for the same crack mode. This is also true for stress intensity factors when *the resulting mode of loading is the same.* Figure 2.10 shows a bending and tensile loading scenario, so we can say that

$$K_I^{total} = K_I^{tension} + K_I^{bending} \tag{2.15}$$

Example 2.3

Derive the stress intensity factor for a semi-infinite plate subjected to a remote tensile stress, σ_r with a center through crack subjected to an internal pressure such that the crack opening stress is equal to the remote stress, σ_r.

The pressurized crack simultaneously subjected to hoop stresses can be decomposed as shown in Figure 2.11. Let us designate the stress intensity factor for the overall case as K_I, and the stress intensity factor for the component cases (a, b, c, etc.) as K_I^i where i = a, b, c, etc.

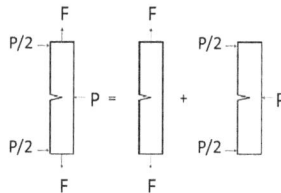

FIGURE 2.10   Stress intensity factor superposition for bending and tension.

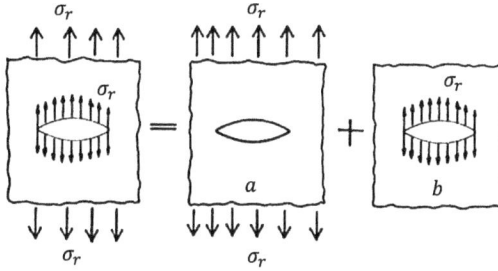

FIGURE 2.11 Decomposition of pressurized center through crack with tensile loading.

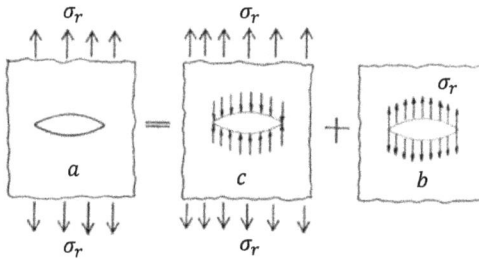

FIGURE 2.12 Decomposition of center through crack in a semi-infinite plate under tensile loading.

We can then write

$$K_I = K_I^a + K_I^b$$

What to make of K_I^b? Consider the following case. A remotely loaded center crack geometry can be decomposed into a set of two center crack geometries which have loading conditions, that when added, result in the canceling of the crack line loadings. This is shown in Figure 2.12.

$$K_I^a = K_I^c + K_I^b$$

For the case (c), where closing stresses (with magnitude σ_r) are present, the crack is clamped closed, that is $K_I^c = 0$, therefore

$$K_I^a = 0 + K_I^b$$

$$K_I^a = K_I^b$$

$$\Rightarrow K_I = 2K_I^a$$

$$K_I = 2\sigma_r\sqrt{\pi a}$$

PROBLEMS

Problems 2.1 through 2.6 lack certain details. Write down *three appropriate questions* to ask prior to applying the equations provided in this chapter to obtain solutions.

2.1. A certain material is subjected to a stress of 300 MPa. What is the maximum stress occurring in an elliptical internal crack in this material with a length of 2.6×10^{-2} mm and a radius of curvature of 3.5×10^{-4} mm?

2.2. What is the maximum length of an elliptical external crack with a radius of curvature of 2.4E-4 mm? The material is subjected to a stress of 450 MPa and the maximum allowable stress in the material is 4,000 MPa.

2.3. A brittle material has a specific surface energy of 1.1 J/m and a modulus of elasticity of 70 GPa. It has an external crack with a length of 0.2 mm. What is the maximum amount of stress that will not cause crack propagation? A certain material is subjected to a stress of 300 MPa. What is the maximum stress occurring in an elliptical internal crack in this material with a length of 2.6×10^{-2} mm and a radius of curvature of 3.5×10^{-4} mm?

2.4. What is the maximum length of an elliptical external crack with a radius of curvature of 2.4E-4 mm? The material is subjected to a stress of 450 MPa and the maximum allowable stress in the material is 4,000 MPa.

2.5. A brittle material has a specific surface energy of 1.1 J/m and a modulus of elasticity of 70 GPa. It has an external crack with a length of 0.2 mm. What is the maximum amount of stress that will not cause crack propagation?

2.6. A brittle material has a specific surface energy of 0.89 J/m and a modulus of elasticity of 78 GPa. It has an internal crack with a length of 4.5E-4 mm. The material is subjected to 300 MPa of stress. The material must have a safety factor of 2.5. Will the stress in the material exceed the maximum allowable stress to avoid crack propagation?

2.7. Use superposition of simpler cases to derive the stress intensity factor–crack length relationship for an edge cracked plate simultaneously subjected to tensile and bending loads. The bending moment is applied in a manner that places the crack in tension. See Appendix B and Table 2.1.

2.8. Use the multiplicity property of the geometry factor to derive the stress intensity factor–crack length relationship for a crack emanating from a hole in an infinite plate subjected to tensile stress that is perpendicular to the crack direction.

FUNDAMENTALS OF LINEAR ELASIC FRACTURE MECHANICS: INTERMEDIATE

OBJECTIVES

After studying the second part of the chapter (Sections 2.7–2.12), the student should be able to:

1. Understand what fracture toughness is and be able to describe how it varies with thickness.

2. Be able to solve an elementary LEFM problem.
3. Understand the behavior of the singularity zone and plastic zone and how they are influenced by the plane strain/plane stress assumption.
4. Understand and apply the concept of effective crack length in the development of solutions for relevant LEFM problems.
5. Be able to explain how fracture toughness typically varies with temperature, strain rate, and strength for metals.

2.7 FRACTURE TOUGHNESS

Irwin [4] postulated that a crack would propagate when the SIF exceeds some critical value. This critical value is called the fracture toughness of the material and is typically denoted K_c. A material's fracture toughness can be measured experimentally; however, it does vary with a material's thickness within a certain range. Figure 2.13 shows the variation of K_c with thickness. The material fracture toughness decreases as the material thickness increases up to a certain point. Beyond this thickness value, the fracture toughness remains constant and therefore can be regarded as a material property.

Recall the discussion of the plane stress–plane strain conditions in Section 2.3. For cracks loaded in tension, the extreme stresses at the crack tip will result in plastic

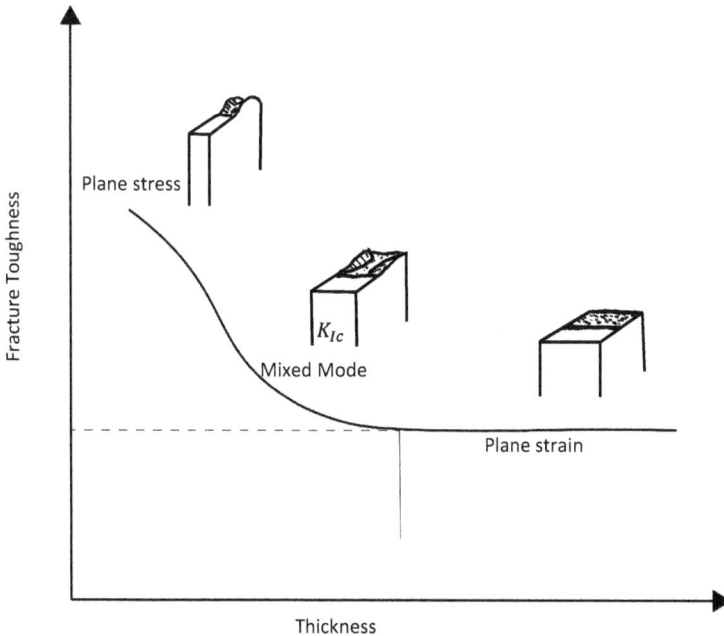

FIGURE 2.13   Fracture toughness variation with thickness.

strains if these strains are physically allowable (i.e., there are no constraints on the material which prevent deformation). When plastic strains are allowed, failure will occur in a ductile manner. This will be the case for thin bodies. The material within the crack tip stress field close to these free surfaces can deform laterally as the crack surfaces separate. As failure is ductile, a 45^0 shear lip develops near each free surface. These shear lips will constitute most of the crack surface. This is seen on the thin body crack image in Figure 2.13.

For thick bodies, the plastic zone beyond the crack tip in the vicinity of the mid-thickness plane is prevented from deforming in the direction of crack propagation, as well as laterally, by the surrounding material. This creates a triaxial stress state in the plastic zone. The shear lips close to the free surfaces will only occupy a small percentage of the crack surface as shown in Figure 2.13. The loading will now approximate a plane strain condition (i.e. no lateral strain) and the crack propagates primarily though brittle failure. The experimentally measured plane strain fracture toughness, K_{Ic}, now remains constant as the specimen thickness increases and can therefore be used as a conservative estimate of the fracture toughness of the material. A curve fit to the graph in Figure 2.13 yields:

$$K_c = K_{Ic}\left(1 + B_k e^{-A_k (B/B_0)^2}\right) \tag{2.16}$$

where K_c is the plane stress fracture toughness, K_{Ic} is the plane strain fracture toughness, B is the material thickness, A_k, B_k are material constants that may be found in material databases, and B_0 is the conventionally accepted value for the thickness at which the plane stress condition becomes a plane strain condition; this is given by

$$B_0 = 2.5\left(\frac{K_{Ic}}{\sigma_{ys}}\right)^2 \tag{2.17}$$

where σ_{ys} is the yield strength. Therefore, a plane strain condition is assumed if $B > B_0$, that is

$$B \geq 2.5\left(\frac{K_{Ic}}{\sigma_{ys}}\right)^2 \tag{2.18}$$

Since K_{Ic} is constant beyond a certain thickness, it is conservatively chosen as *the maximum value that the stress intensity factor can be prior to a crack becoming unstable under Mode I loading.* It is a measure of the material's resistance to brittle failure. Ductile materials therefore generally have a higher fracture toughness than brittle materials. Table 2.4 provides fracture toughness values of selected materials. ASTM standards E1820-18, ASTM E399, E1290, and E1921 provide standard test methods to measure the fracture toughness for a variety of materials. These will be discussed in detail in Chapter 6.

TABLE 2.4

Plane Strain Fracture Toughness Values for Selected Materials

Material	$K_{Ic}\left(\text{MPa}/\sqrt{m}\right)$
Aluminum alloys	22–35
Titanium alloys	14–120
Low carbon steel alloys	40–80
Concrete	0.35–0.45
Wood	5–9
Glass	0.6–0.8

Example 2.4

A 7075-T6 aluminum beam is subjected to a bending load as shown in Figure 2.14(a). The length of the beam is 10 m. Its rectangular cross section is 0.3 m high and 0.06 m thick. The forging direction is along the longitudinal axis. What is the maximum value of the load, P, that can be applied so that failure does not occur (assume the bar fails when yielding occurs)?

If the bar had a 0.03 m through crack as shown in Figure 2.14(b), use LEFM to determine the maximum load, P, that can be applied to the bar so that the crack does not propagate. $K_{Ic} = 29\,\text{MPa}\sqrt{m}$, $\sigma_{ys} = 501$ MPa. Assume a safety factor of 3. The stress intensity relationship for a uniform beam under bending load with a central crack is provided in Appendix C.

The first part of the problem requires a standard mechanics of materials approach. The maximum tensile stress occurs on the bottom surface in the center of the bar. This can be found from

$$\sigma = \frac{Mc}{I}$$

where M is the bending moment, c is the distance from the neutral axis (i.e. $c = 0.15$ m) and I is the area moment of inertia of the cross-section about the neutral axis. At the center of the beam, M is given by:

$$M = \frac{P}{2}\times 5 = \frac{5P}{2}$$

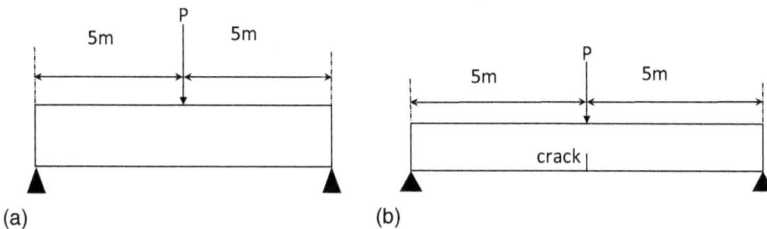

(a) (b)

FIGURE 2.14 (a) Beam with bending load, (b) cracked beam with bending load.

I about the neutral axis of the cross-section is found from:

$$I = \frac{0.06 \times 0.3^3}{12} = 1.35 \times 10^{-4} \, \text{m}^4$$

The maximum stress is therefore:

$$= \frac{\frac{5P}{2} \times 0.15}{1.35 \times 10^{-4}} = \frac{P}{36} \times 10^5 \, \text{Pa}$$

The maximum allowable stress is $\sigma_{ys}/3 = 167$ MPa, therefore the maximum load, P, can be found by:

$$\frac{P}{36} \times 10^5 = 167 \times 10^6, P \approx 60.1 \, \text{KN}$$

FRACTURE MECHANICS APPROACH

For a beam undergoing three-point bending, the stress intensity factor can be found in Appendix C. The stress intensity expression is:

$$K_I = \frac{P}{B\sqrt{W}} \frac{3\left(\frac{S}{W}\right)\sqrt{\frac{a}{W}}}{2\left(1+2\left(\frac{a}{W}\right)\right)\left(1-\left(\frac{a}{W}\right)\right)^{3/2}}$$

$$\left[1.99 - \frac{a}{W}\left(1-\frac{a}{W}\right)\left\{2.15 - 3.93\left(\frac{a}{W}\right)+2.7\left(\frac{a}{W}\right)^2\right\}\right]$$

where W is the beam height, B is the beam thickness and S is the beam length. Substituting $B = 0.06$ m, $\frac{a}{W} = \frac{0.03}{0.3} = 0.1, S = 10$ m, we obtain:

$$K_I = 858.87P$$

The crack will become unstable when $K_I = K_{Ic}$. Using a safety factor of 3, the allowable fracture toughness is $\frac{29}{3}$ MPa$\sqrt{\text{m}}$, therefore:

$$858.87P = \frac{29}{3} \times 10^6; P \approx 11.3 \, \text{kN}$$

The existence of the 0.03 m crack at the center of the beam reduces the maximum allowable load capacity by approximately 81%.

2.8 THE SINGULARITY ZONE AND THE PLASTIC ZONE

Figure 2.15 shows the *local* stress distribution as we move away from the crack tip for $\theta = 0$ and Mode I (see Table 2.1). For this case, the remote stress, σ_{yy}, in the vicinity of the crack tip is given by

$$\sigma_{yy} = \frac{K_I}{\sqrt{2\pi r}} \tag{2.19}$$

Equation (2.19) predicts that a stress singularity exists at the crack tip. In the physical world, such a singularity cannot occur as the material will yield locally and plastically deform. The stresses close to the crack tip are still however very high and decrease as we move further away from the tip, eventually tapering off to the magnitude of the remote stress, σ_∞. The distance over which the $1/\sqrt{r}$ term dominates the stress behavior is called the *singularity-dominated zone*. If such a zone exists, then the fracture is K-controlled and an LEFM approach is valid.

Not all the stresses in the singularity zone will necessarily be above yield. The area surrounding the crack tip where the stresses are above yield is referred to as the *plastic zone*. Any plastic behavior that occurs near the crack tip serves to reduce the local stresses [5]. These reduced stresses are not dependent on the $1/\sqrt{r}$ term. Therefore, as the plastic zone increases in size, the $1/\sqrt{r}$ term becomes less dominant, rendering LEFM less accurate.

Once the plastic zone grows to the size of the singularity zone, the fracture is no longer K-controlled and LEFM will not be valid. Quantitative rules for the validity of LEFM are provided in the subsequent section. Figure 2.16 shows the effect of the plastic zone of length r_y on the stress distribution in the vicinity of the crack tip.

Several successful correction methods that account for crack tip plasticity have been devised. The Irwin method, one of the simpler accepted methods, is presented below. Irwin's approach determines the size of the plastic zone, r_{yo}, by assuming that yielding occurs when the local stress at the crack tip equals the yield strength of the

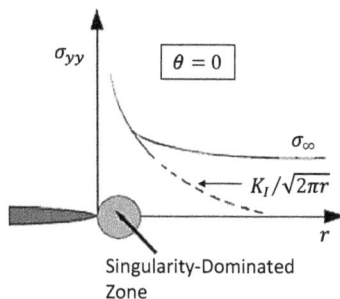

FIGURE 2.15   Stress normal to crack plane Mode I loading.

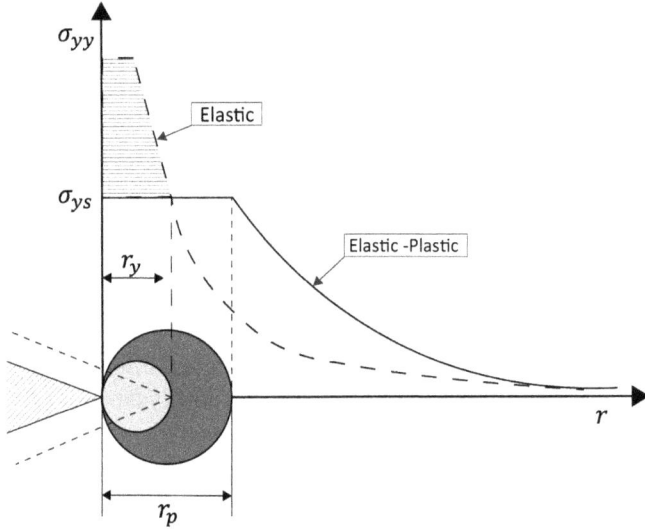

FIGURE 2.16   Plastic zone size and stress distribution at crack tip (Mode I).

material under plane stress conditions. Therefore, we can substitute $\sigma_{yy} = \sigma_{ys}$ in Eqn. (2.19) to obtain

$$r_{y\sigma} = \frac{1}{2\pi}\left(\frac{K_I}{\sigma_{ys}}\right)^2 \qquad (2.20)$$

Equation (2.20) is a first-level estimate for the length of the plastic zone under a *plane stress condition*. It does not consider the stress redistribution due to the high stresses from the elastic analysis. A higher-level analysis that considers this redistribution by enforcing equilibrium results in doubling the plastic zone length; the second level estimate for this length with the plane stress condition, denoted by r_p in Figure 2.16, is therefore

$$r_p = 2r_{y\sigma} = \frac{1}{\pi}\left(\frac{K_I}{\sigma_{ys}}\right)^2 \qquad (2.21)$$

For the *plane strain condition*, the triaxial stress field suppresses the plastic zone size by a factor of 3, to yield a plastic zone size, $r_{y\varepsilon}$:

$$r_{y\varepsilon} = \frac{1}{6\pi}\left(\frac{K_I}{\sigma_{ys}}\right)^2 \qquad (2.22)$$

Substitution of $K_I = K_{Ic}$ in Eqns (2.20) and (2.22) provides an upper bound on the plastic zone size for the plane stress and plane strain condition, respectively.

2.9   EFFECTIVE FRACTURE TOUGHNESS

The plastic zone is incorporated into LEFM by introducing an effective crack length (see Figure 2.16), which is the sum of the plastic zone length and the original crack length, a, that is

$$a_{eff} = a + r_y \tag{2.23}$$

where the magnitude of r_y depends on whether a plane stress ($r_y = r_{yo}$) or plane strain condition ($r_y = r_{ye}$) can be approximated.

An effective stress intensity, K_{eff}, can then be obtained from a_{eff} using the relevant closed form solutions such as those in Table 2.2. Note that each new K_{eff} will result in a new a_{eff}, therefore the method may be repeated until successive K_{eff} values converge. It's important to note that the geometry correction factor Y will also be a function of a_{eff} as shown in Eqn (2.24). As a_{eff} increases, finite width effects become more relevant and changes in Y should be considered in the analysis.

$$K_{eff} = Y\left(a_{eff}\right)\sigma\sqrt{\pi a_{eff}} \tag{2.24}$$

In some cases, an iterative process is not needed as a closed form solution may be derived. In the case of a thin infinite plate with a central through crack, we can substitute for r_y from Eqn (2.20) ($K_I = K_{eff}$) into Eqn (2.23) and apply the resulting a_{eff} to Eqn. (2.24), which yields

$$K_{eff} = \frac{\sigma\sqrt{\pi a}}{\sqrt{1 - \frac{1}{2}\left(\sigma/\sigma_{ys}\right)^2}} \tag{2.25}$$

Dugdale [6] and Barenblatt [7] derived a closed form solution by superimposing the LEFM solution for a through crack under remote tension and a through crack with closure stresses at the crack tip equal to the material yield strength. This model is called the *strip-yield* model, resulting in

$$K_{eff} = \sigma\sqrt{\pi a \sec\left(\frac{\pi\sigma}{2\sigma_{ys}}\right)} \tag{2.26}$$

Burdekin and Stone [8] later improved on this closed form expression by using a complex stress function with the strip-yield model, yielding :

$$K_{eff} = \sigma_{ys}\sqrt{\pi a}\left[\frac{8}{\pi^2}\ln\left(\sec\frac{\pi\sigma}{2\sigma_{ys}}\right)\right]^{1/2} \tag{2.27}$$

2.9.1 CRITERIA FOR LEFM VALIDITY

Irwin showed that LEFM will no longer be valid when the plastic zone is the size of the singularity zone. The size of the singularity zone is determined by the distance from the crack tip, r, at which Eqn. 2.19 no longer correctly predicts the local stresses. When $r \approx \dfrac{1}{50} a$, Eqn. 2.19 underestimates σ_{yy} by 2% and overestimates σ_{xx} by 20%. This distance is therefore conventionally chosen to define the outer limit of the singularity-dominated zone.

For the more conservative *plain strain* condition:

$$r_y = \frac{1}{6\pi}\left(\frac{K_I}{\sigma_{ys}}\right)^2$$

$$\Rightarrow \frac{1}{6\pi}\left(\frac{K_I}{\sigma_{ys}}\right)^2 \le \frac{1}{50} a$$

or

$$a \ge \frac{25}{3\pi}\left(\frac{K_I}{\sigma_{ys}}\right)^2 \tag{2.28}$$

ASTM's version (ASTM E399) of the necessary geometrical limits for the validity of LEFM is shown in Eqn (2.29). This is a slight modification of Eqn (2.28) where the constant is reduced from 2.65 to 2.5. Figure 2.17 shows the geometric parameters involved. As the crack grows, the true remote stress in the remaining ligament (net section stress) may be considerably higher than the predicted K-dominated stress, so the minimum value is also imposed on the ligament $W - a$, as well. When Eqn (2.29) is satisfied, the condition is referred to as *small scale yielding* because the size of the plastic zone is small relative to the singularity-dominated zone.

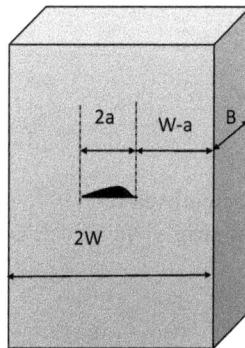

FIGURE 2.17 Relevant geometrical features for LEFM applicability limits (ASTM)

$$a, W - a \geq 2.5 \left(\frac{K_I}{\sigma_{ys}} \right)^2 \qquad (2.29)$$

The standard also suggests that the specimen thickness, B, for plane strain fracture toughness testing (tension), satisfies $2 \leq \dfrac{W}{B} \leq 4$, and recommends a nominal value of $\dfrac{W}{2}$ for both a and B. If these nominal values are used, we can conservatively say that $B \geq 2.5 \left(\dfrac{K_I}{\sigma_{ys}} \right)^2$ is also an appropriate minimum thickness criterion for the plane strain condition.

Concept Challenge 2.4

How would strain hardening affect the size of the plastic zone?

Example 2.5

A center cracked panel with a 1 in crack is subjected to a uniform tensile stress of 22 ksi. The toughness is $50 \, \text{ksi}\sqrt{\text{in}}$, the yield stress is 70 ksi, the plate width is 14 in, and the plate thickness is 1 in. Is LEFM applicable? Should a plane stress or plane strain condition be used? What is the size of the plastic zone? What is the LEFM Safety Factor?

Does LEFM apply? We can apply Eqn (2.29) to check if LEFM is applicable. We therefore need to determine the stress intensity factor K_I:

$$K_I = Y \sigma \sqrt{\pi a}$$

For a center cracked specimen $Y \approx 1$

$$K_I = (1) 22 \sqrt{\pi (0.5)} = 27.6 \, \text{ksi}\sqrt{\text{in}}$$

LEFM is applicable if

$$a, W - a \geq 2.5 \left(\frac{K_I}{\sigma_{ys}} \right)^2$$

$$2.5 \left(\frac{K_I}{\sigma_{ys}} \right)^2 = 2.5 \left(\frac{27.6}{70} \right)^2 = 0.39 \, \text{in}$$

$$W - a = 6.5 \, \text{in} \geq 0.39 \, \text{in}$$

$$a = 0.5 \, \text{in} \geq 0.39 \, \text{in}$$

Therefore, LEFM is applicable.

Plane stress or plane strain?

The minimum thickness, B, for a plane strain condition is shown in Eqn (2.18). Since the plate thickness, 1 in, is greater than 0.39 in, a plane strain condition can be used.

The length of the plastic zone, r_p, is given by Eqn. (2.22), therefore

$$r_p = \frac{1}{6\pi}\left(\frac{K_I}{\sigma_{ys}}\right)^2 = \frac{1}{6\pi}\left(\frac{27.6}{70}\right)^2$$

$$r_p = 0.008 \text{ in.}$$

The LEFM Safety Factor (Chapter 1) is given by

$$\frac{K_{Ic}}{K_I} = \frac{50}{27.6} = 1.81$$

Example 2.6

An aluminum alloy of yield stress 400 MPa is tested in thin wide sheets. For a central crack of length 20 mm the fracture stress (tensile) is observed to be 200 MPa. Determine the fracture toughness for the alloy using (a) a first iteration for the plasticity correction and (b) the closed form solution, i.e. Eqn (2.25).

We start with the calculation assuming no plastic zone, i.e.

$$K_{Ic} = 200\sqrt{0.01\pi} = 35.45 \text{ MPa}\sqrt{m}$$

Since the plate is thin, we can assume the plane stress condition. The additional crack length r_y due to the plastic zone is found from Eqn (2.20) using the K_{Ic} calculated above, i.e.

$$r_y = \frac{1}{2\pi}\left(\frac{K_{Ic}}{\sigma_{ys}}\right)^2$$

$$= \frac{1}{2\pi}\left(\frac{35.45}{400}\right)^2$$

$$r_y = 1.25 \text{ mm}$$

The effective crack length now becomes

$$a_{eff} = a + r_y$$

$$a_{eff} = 0.01 + 0.00125 = 0.01125 \text{ m}$$

Therefore

$$K_{eff} = 200\sqrt{0.01125\pi} = 37.6 \text{ MPa}\sqrt{m}$$

The closed form solution yields

$$K_{eff} = \frac{\sigma\sqrt{\pi a}}{\sqrt{1-\left(\frac{1}{2}\right)\left(\sigma/\sigma_{ys}\right)^2}} = \frac{200\sqrt{0.01\pi}}{\sqrt{1-\left(\frac{1}{2}\right)\left(200/400\right)^2}} = 37.9 \text{ MPa}\sqrt{m}$$

2.10 FRACTURE TOUGHNESS AND OTHER PROPERTIES

2.10.1 FRACTURE TOUGHNESS AND STRENGTH

The relationship between fracture toughness and strength (tensile, yield) depends on the class of material. For metals and most semi-crystalline polymers, fracture toughness is inversely proportional to tensile strength. Metallic bonding allows for the occurrence of dislocations, that is the mechanism responsible for ductile behavior.

In semi-crystalline polymers, the chains of atoms can slip past each other under an applied load. The greater the resistance to dislocations (in the case of metals) or chains slipping past each other (in the case of polymers), the greater the tensile strength. So, the primary mechanisms responsible for increasing tensile strength in metals and polymers will also decrease fracture toughness. Yield strength may be increased by dispersion additions, strain hardening, or solid solution strengthening. Any increase in yield strength decreases plastic zone size, decreasing energy absorption. Therefore, fracture toughness also decreases with an increase in yield strength. Figure 2.18 is a schematic diagram that shows the relationship between fracture toughness and strength for metals and most semi-crystalline polymers. A similar inverse relationship between fracture toughness and tensile strength exists for most ceramics. Ceramics typically have high compressive strengths, low tensile strengths and low fracture toughness, compared to metals.

The fracture toughness and tensile strength of many geomaterials such as cohesive soil, frozen soil, soft rock, and hard rock have been found to be linearly correlated [9].

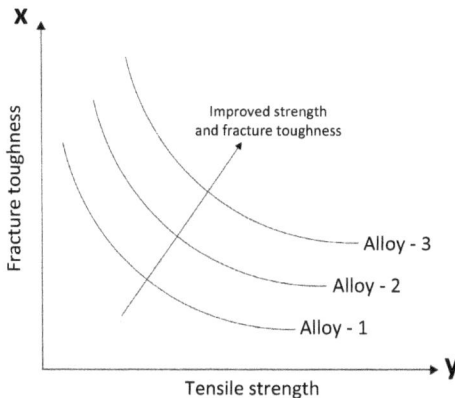

FIGURE 2.18 Fracture toughness vs. tensile strength for metals and most semi-crystalline polymers.

2.10.2 Fracture Toughness vs. Temperature

An increase in temperature within a material implies that the kinetic energy of the atoms increases, that is the amplitude and velocity of the atomic vibrations will increase. Therefore dislocations, the mechanism responsible for plastic deformation, will occur more frequently and move at a faster rate. The material therefore tends to have increased ductility. Generally, as temperature increases, the fracture toughness of a material will increase.

2.10.3 Fracture Toughness vs. Strain Rate

The influence of the strain rate on fracture toughness is complex. Within different ranges, new phenomena may occur to change the nature of the relationship (e.g. cleavage fracture vs. ductile fracture). Additionally, whether the material is ductile or brittle may change the response. Consequently, the response may be different if the temperature is above or below the material's ductile to brittle transition temperature.

For rates lower than $10^3 s^{-1}$, dislocations are thermally assisted across short range barriers. This thermal softening is more dominant within brittle materials. It can be expected that the fracture toughness increases with an increasing strain rate for these materials. For a ductile material at a given temperature, subjected to strain rates within the aforementioned range, an increase in strain rate in the elastic loading region generally reduces the fracture toughness as the thermal softening is less dominant [10, 11]. At very high strain rates ($>10^4 s^{-1}$), the behavior is controlled by the interaction of dislocations with either phonons or electrons; adiabatic heating may also occur. These phenomena result in a fracture toughness that is proportional to the strain rate. Table 2.5 provides the strain and loading rates one might encounter in industry applications.

A smaller grain size results in more grain boundary area; these grain boundaries tend to inhibit dislocation movement, leading to an increase in yield strength. An increase in yield strength is typically accompanied by a decrease in fracture toughness; however, this does not happen in this case. The larger number of grain boundaries also results in a dislocation pile up at each boundary which increases the stress at the boundary. A crack needs more energy to traverse the grain boundaries due to the increased grain boundary area and to overcome compressive stress fields generated from dislocation pile up. Generally, fracture toughness increases with a reduction in grain size.

TABLE 2.5
Typical Strain and SIF Loading Rates in Industry Applications [12, 13]

Application	Strain Rate (s^{-1})	SIF Loading Rate $\left(MPa\sqrt{m}\ s^{-1} \right)$
Storage tanks, pressure vessels	10^{-6}–10^{-4}	10^{-2}–10
Bridges, earth-moving cranes	10^{-2}–10^{-1}	10^1–10^3
Earthquake loading, marine collisions	10^{-1}–10^1	10^2–10^4
Land transport, aircraft undercarriage	10^1–10^3	10^3–10^6
Explosion, ballistics	10^4–10^6+	10^7–10^{10}+

2.11   INDUSTRY APPLICATIONS

2.11.1   HYDRAULIC PROOF TEST TO DETERMINE K_{Ic}

Cylindrical pressure vessels built in the 1960s are being used at a current industrial facility. The only material information the company can recover for these vessels; is that the yield strength is 1000 MPa. In order to estimate K_{Ic}, they adopt the following procedure. An artificial longitudinal surface crack is imparted in the material with the expectation that this will serve as the fracture initiation point when the vessel is subjected to increasing internal pressure. A hydraulic proof test is then performed, and the vessel undergoes unstable fracture at an applied hoop stress of 600 MPa. An inspection after the test reveals that the artificial surface crack served as the location of fracture initiation (as intended) and that it grew to 3 mm deep by 10 mm wide just prior to fracture. How can the fracture toughness be determined based on this data?

Figure 2.7 provides the aspect ratio vs. flaw size for various applied stress levels (normalized by the yield stress). Since the aspect ratio, fracture, and yield stress are known, this figure can be used to determine the flaw shape parameter, Q.

The aspect ratio is $\dfrac{a}{2c} = \dfrac{3}{10} = 0.3$, the stress ratio $\dfrac{\sigma}{\sigma_y} = \dfrac{600}{1000} = 0.6$.

Examining Figure 2.7, we see that the flaw shape parameter $Q = 0.15$.
We can now use Eqn (2.11) to estimate the fracture toughness

$$K_{Ic} = 1.12\sigma\sqrt{\pi\frac{a}{Q}} = 1.12 \times 600 \times 10^6 \times \sqrt{\frac{\pi \times 3 \times 10^{-3}}{0.15}}$$

$$K_{Ic} = 168.45\,\text{MPa}\sqrt{\text{m}}$$

Concept Challenge 2.5

What difference would it have made had the artificial crack been oriented at some angle relative to the longitudinal axis?

2.11.2   FRACTURE TOUGHNESS VS. TENSILE STRENGTH

2¼ Cr-1Mo steel is used in the oil and gas, energy, construction, and automotive industries because of its corrosion resistance and high temperature and tensile strength. A spherical storage tank (radius 10 m, thickness 5 cm) is to be fabricated from large sheets of 2¼ Cr-1Mo steel. *A design stress level of half the tensile strength is proposed.* For this scenario, assume flaw sizes of 15 mm or greater can be detected by feasible non-destructive tests (NDTs). It has been suggested that the steel could be heat treated to a higher tensile strength level, which would in turn yield a higher pressure rating. Within the expected service temperatures, the current grade has a tensile strength of 500 MPa, and a candidate replacement grade has a 2,000 MPa

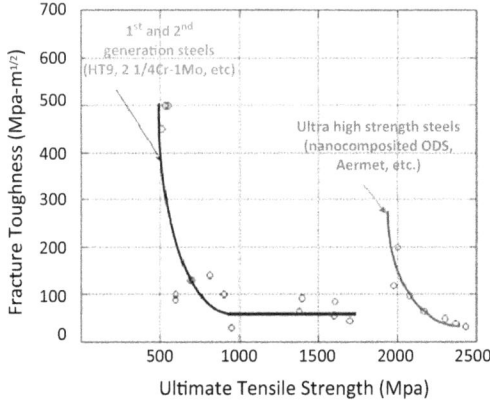

FIGURE 2.19 Fracture toughness vs. yield strength for various steels [14].

strength level. Figure 2.19 shows the relationship between fracture toughness and tensile strength for this steel among others. Use fracture mechanics and Figure 2.19 to examine whether the higher pressure rating for the tank is feasible. Assume plane strain conditions.

The stress-based approach would yield a design pressure based on the tensile strength. For a spherical vessel subjected to internal pressure, the stress state is biaxial where the tensile stress for both axes is given by

$$\sigma = \frac{Pr}{2t}$$

The contribution of the tensile stress parallel to the crack direction may be neglected.

Since r and t are constant, by inspection we see that using the higher tensile strength version would increase the internal pressure capacity by a factor of 4.

If a small crack exists in the thin-walled pressure vessel, the local vicinity around the small crack can be treated as an infinite plate under tension, therefore

$$K_{Ic} = \sigma\sqrt{\pi a_{crit}}$$

From Figure 2.19, for the 2¼ Cr-1Mo with a 500 MPa ultimate strength, $K_{Ic} = 400 \text{ MPa}\sqrt{m}$

$$400 = 250 \times \sqrt{\left(\pi \times a_{crit}\right)}$$

$$a_{crit} = 81.5 \text{ cm}$$

$$\text{Total flaw size} = 2a_{crit} = 163 \text{ cm}$$

If we choose to use ultra-high strength steel with a 2,000 MPa ultimate tensile strength, we see from Figure 2.19 that $K_{Ic} \approx 150\ \mathrm{MPa}\sqrt{m}$, therefore

$$150 = 1000 \times \sqrt{\left(\pi \times a_{crit}\right)}$$

$$a_{crit} = 7.2\ \mathrm{mm}$$

$$\text{Total flaw size} = 2a_{crit} = 14.4\ \mathrm{mm}$$

While the higher strength steel would provide substantially increased pressure capacity, it is not feasible from a fracture mechanics perspective. The flaw size of 14.4 mm prior to unstable crack growth is not detectable with the available NDT technology. Additionally, implementing a design that depends on a crack being found by NDT methods when it is at or close to its critical length is an extremely poor and dangerous design decision.

2.12 REVIEW AND MATERIAL ASSUMPTIONS

The relationships described are valid for linear elastic homogeneous materials. LEFM can be extended to composites; however the fracture mechanics of composites is beyond the current scope of this chapter. LEFM seeks to determine the SIF for a given service loading and geometry. The SIF is then compared to a critical SIF called the fracture toughness for that material. If the fracture toughness value has been exceeded within a material for some existing crack size, this is an indication that the crack will be an unstable one.

1. The SIF varies with loading and geometry. SIF closed-form solutions are listed in many texts, reports, and online. SIFs for complex loading and geometry may be obtained through computational methods such as finite element analysis.
2. Fracture toughness values will be different depending on the orientation of the crack with grain direction. Care should be taken to ensure that the fracture toughness chosen is for the same geometry, loading, and crack orientation with respect to grain direction. This is discussed further in Chapter 6.
3. Experimentally measured fracture toughness, K_{Ic}, decreases with increasing thickness; however, after a certain thickness it remains constant. At this thickness it is called the plane strain fracture toughness, K_{Ic}, which is the value used as the maximum limit for the stress intensity factor in fracture mechanics design problems.
4. Stress, displacement, strain, and stress intensity solutions from individual loads may all be added together for the LEFM problem that involves multiple loads if the crack modal loading is the same.
5. There is an inelastic region near the crack tip that serves to increase the effective crack length; this results in corrections to the stress intensity factor and the introduction of an effective stress intensity factor.

PROBLEMS

2.9. Use Eqns (2.20, 2.23, and 2.24) to prove Eqn (2.25).

2.10. A material has a plain strain fracture toughness of 30 MPa$\sqrt{\text{m}}$. The stress in the material is 250 MPa. There is an external crack of length 3.8 mm. What is the value of the geometric factor Y?

2.11. A material has a plain strain fracture toughness of 45 MPa $\sqrt{\text{m}}$. The stress in the material is 300 MPa. Assume that the geometric factor $Y = 0.86$. What is the maximum external crack length?

2.12. For an internal crack in an infinite plate loaded in tension oriented at an angle β relative to the direction, examine the expressions for K_I and K_{II}. Show using Mohr's circle why these equations necessarily follow from the K_I solution at $\beta = 0$.

2.13. A large plate $(2a \ll w)$ containing a central crack 30 mm long is subjected to a tensile stress perpendicular to the crack orientation. The crack growth rate is 12 mm/month and the fracture is expected in five months. Given that $K_{IC} = 30\,\text{MPa}\sqrt{\text{m}}$, calculate the fracture stress.

2.14. A tensile sample of width 10 mm contains an internal crack of length 0.3 mm. When loaded in tension the crack suddenly propagates when the stress reaches 450 MPa. What is the fracture toughness K_{Ic} of the material of the sample? If the material has a modulus E of 200 GPa, what is its critical strain energy release rate G_c?

2.15. Consider a steel plate $\left(K_{Ic} = 45\,\text{MPa}\sqrt{\text{m}}, \sigma_{YS} = 1200\,\text{MPa} \right)$ with a through thickness edge crack loaded in tension. The configuration is that of a single-edge notched tension (SENT) specimen as shown in Appendix C. The plate width is 100 mm, thickness 8 mm. The service load along the axis perpendicular to the crack (and in the same plane) is expected to be 240 kN. Will failure occur if the crack length is 30 mm? Why/why not?

2.16. What yield strength would the material in problem 2.10 need to have in order to ensure that LEFM is applicable?

2.17. Draw a flow chart for a computer program that uses an iterative process to determine the effective stress intensity factor K_{eff} according to the Irwin model. Apply this to your program of choice (e.g. MATLAB) and use to solve the following problem. Consider a thin steel plate of width 50 mm with a central crack of length 15 mm subjected to a stress of 400 MPa normal to the crack plane. Assume the yield stress of the material is 3,000 MPa. Plot the normal stress ahead of the crack tip according to the Irwin model.

2.18. A cylindrical pressure vessel made of material of fracture toughness 82 MPa$\sqrt{\text{m}}$ contains a large embedded circular crack of length $2a = 10$ mm. The radius of the cylinder is 10 m and its wall thickness is 2 cm. What is the maximum pressure the vessel can withstand before rupture?

2.19. Silver brazing is a joining process whereby a non-ferrous filler metal alloy is heated to melting temperature and distributed between two or more close-fitting parts. A copper rod and a steel rod are brazed end-to-end as shown in Figure 2.20. Both have a rectangular cross-section. Braze failure is known to occur at 400 MPa. The steel and copper rod sections have yield strengths of 320 MPa and 195 MPa, respectively, and plane-strain fracture-toughness values of $80\,\text{MPa}\sqrt{\text{m}}$ and $95\,\text{MPa}\sqrt{\text{m}}$, respectively. Assume each rod could contain an elliptical surface flaw 1.0 cm deep and 3.3 cm wide that is oriented normal to the stress axis. If the rod/braze assembly is loaded in tension at 192 MPa perpendicular to the joint plane, will failure first occur in the braze joint or in the steel or in the copper?

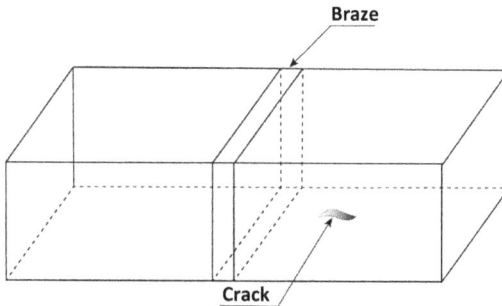

FIGURE 2.20 Failure of brazed joint.

2.20. A thin-walled cylindrical pressure vessel with closed ends is subjected to an internal pressure, P. The radius of the vessel is 450 mm and the wall thickness is 15 mm. The plane strain fracture toughness of the vessel is $K_{Ic} = 77\,\text{MPa}\sqrt{\text{m}}$. Determine the pressure at which the vessel will fracture if:

(a) A 20 mm long through-thickness crack exists in the vessel along the longitudinal axis of the vessel.

(b) A 20 mm long through-thickness crack exists in the vessel in the circumferential (hoop) direction.

REFERENCES

[1] E. Kirsch, "Die Theorie der Elastizität und die Bedürfnisse der Festigkeitslehre," *Zeitschrift des Vereines deutscher Ingenieure*, vol. 42, pp. 797–807, 1898.

[2] C. Inglis, "Stresses in a Plate due to the Presence of Cracks and Sharp Corners," *Trans Inst Nav Archit*, vol. 55, pp. 219–230, 1913.

[3] A. Griffith, "The phenomena of rupture and flow in solids," *Philosophical Transactions*, vol. Series A, no. 221, pp. 163–198, 1920.

[4] G. Irwin, "Analysis of stresses and strains near the end of a crack traversing a plate," *Transactions ASME, Journal of Applied Mechanics*, vol. 24, pp. 361–364, 1957.

[5] P. Charalambides and R. McMeeking, "Finite element method simulation of crack propagation in a brittle microcracking solid," *Mechanics of Materials*, vol. 6, no. 1, pp. 71–87, 1987.

[6] D. Dugdale, "Yielding in steel sheets containing slits," *Journal of the Mechanics and Physics of Solids*, vol. 8, pp. 100–104, 1960.

[7] G. Barenblatt, "The Mathematical Theory of Equilibrium Cracks in Brittle Fracture," *Advances in Applied Mechanics*, vol. 7, pp. 55–129, 1962.

[8] F. Burdekin and D. Stone, "The crack opening displacement approach to fracture mechanics in yielding materials," *Journal of Strain Analysis*, vol. 1, pp. 145–153, 1966.

[9] J. Wang, J. Zhu, and H. Zhang, "Experimental study on fracture toughness and tensile strength of a clay," *Engineering Geology*, vol. 94, no. 1–2, pp. 65–75, 2007.

[10] C. S. Wiesner and H. MacGillivray, "Loading Rate Effects on Tensile Properties and Fracture Toughness of Steel," in *1999 TAGSI Seminar - Fracture, Plastic Flow and Structural Integrity*, Cambridge UK, 1999.

[11] A. Krausz and K. Krausz, "The Constitutive Law of Deformation Kinetics," in *Unified Constitutive Laws for Plastic Deformation*, San Diego, Academic Press, Inc., 1996, pp. 272–274.

[12] P. Christopher, "Some Proposals for Dynamic Toughness Measurement," in *The Welding Institute and American Society of Metals International Conference on Dynamic Fracture Toughness*, London, 1976.

[13] M. Dawes, "The Relevance of Deformation Rate to Brittle Fracture Initiation," The Welding Institute Research Bulletin, London, 1976.

[14] S. J. Zinkle and J. T. Busby, "Structural materials for fission & fusion energy," *Materials Today*, vol. 12, no. 11, pp. 12–19, 2009.

3 Energy Approaches

3.1 INTRODUCTION

In the previous chapter, the failure of a structure that contained cracks was examined based on the stress field that developed in the vicinity of the crack tip. The theory of failure factored in the geometry of the crack and the remotely applied forces in the form of a stress intensity factor.

In this chapter, we introduce another concept that is not directly based on the localized stress field at the crack tip as the basis for predicting crack advance. Instead we make use of an energy balance approach applied to the entire body for failure prediction. The localized crack tip stresses have not been ignored, but are instead captured by the resulting strain energy they create within the bonds of the material's lattice structure.

This concept was first proposed by Griffith [1] in 1921, many years before the stress intensity approach was introduced, and it is still being used today in many applications including brittle material failure.

3.2 GRIFFITH'S THEORY

A material under tension will elongate in the tensile direction, therefore the inter-atomic distance increases, resulting in stored strain energy. The stored mechanical strain energy, U, within an infinitely wide plate subjected to tensile loading, having a central transverse crack, is given by

$$U = U_0 - \frac{\pi \sigma^2 a^2 B}{E}$$

(3.1)

where U_0 is the initial strain energy. The other available potential energy source for crack formation is the interatomic bond energy, some of which may be "released" on crack formation, resulting in a decrease in the overall strain energy. Griffith postulated that the energy available for crack nucleation and growth derives from work done by external forces and the strain energy that has been released. The atoms on the surface of a material are not bonded to as many nearest neighbor atoms as those on the interior; these unsatisfied bonds result in an excess energy on the surface called the surface energy, γ_s. Surface energy has units of energy per unit area (e.g. J/m²). Since two surfaces are created when a crack develops, the interatomic bond energy, E_{bond}, is therefore twice the surface energy. For a crack of length $2a$ in a material of thickness B, E_{bond} can be expressed as

$$E_{bond} = 2\gamma_s \times 2aB = 4\gamma_s aB$$

(3.2)

DOI: 10.1201/9781003052050-3

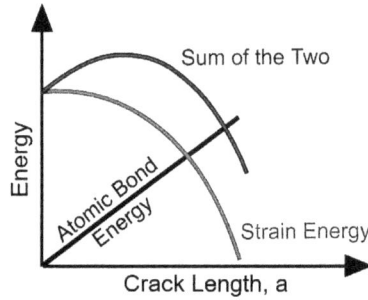

FIGURE 3.1 Energy associated with crack growth.

E_{bond} is therefore a linear function of a. The sum of the interatomic bond energy and the stored mechanical energy is the total energy available for crack growth, which is a quadratic function of a as shown in Figure 3.1.

We observe the parabolic shape of the total energy curve and see that when a is small (i.e. short cracks), $\frac{dE}{da} > 0$, therefore the total energy needed for crack growth increases as crack length increases. As a increases, there is a certain crack length at which $\frac{dE}{da} = 0$; beyond this length $\frac{dE}{da} < 0$. A negative slope implies that the total energy needed for crack growth to occur decreases as crack length increases. This is now an unstable scenario as the crack will grow without any additional energy input. The transition point from stable crack growth to unstable growth is therefore defined by $\frac{dE}{da} = 0$. Griffith used this point, i.e. $\frac{dE}{da} = 0$, to derive Eqn. (3.3), which is very similar to Eqn. (2.6)

$$\sigma_f = \sqrt{\frac{2E\gamma_s}{\pi a}}$$
(3.3)

The proof is left as an exercise for the reader.

Concept Challenge 3.1

Equation (3.3) includes surface energy and plastic flow as contributions to the total energy released per unit area as a crack grows. What other phenomena may contribute to energy dissipation during crack growth?

Equation (3.3) was later modified to account for materials capable of plastic flow, such that

$$\sigma_f = \sqrt{\frac{2E\left(\gamma_s + \gamma_p\right)}{\pi a}}$$
(3.4)

where γ_p is the plastic work per unit area of surface created.

The total energy available for crack formation (see Figure 3.1) is given by $E = U + E_{bond}$. Based on the parabolic shape of the total energy vs. crack length curve, we noted that the point of crack growth instability for a crack surface would occur when $\frac{dE}{da} = 0$. For a crack with surface area A, assuming constant thickness,

$$\frac{dE}{da} = 0 \Rightarrow \frac{dE}{dA} = 0$$

$$\frac{dE}{dA} = 0 \Rightarrow -\frac{dU}{dA} = \frac{dE_{bond}}{dA} \tag{3.5}$$

Equation (3.5) implies that the strain energy per incremental area of crack growth is equal to the interatomic bond energy per unit area at the point of crack instability. The quantity $-\frac{dU}{dA}$ represents a "rate" (per incremental increase in crack area) at which the strain energy is released, resulting in crack growth (i.e. new surface formation). This quantity is called the strain energy release rate and given the symbol G.

Since

$$E_{bond} = 4\gamma_s aB = 2 \times (2aB)\gamma_s = 2A\gamma_s$$

$$\frac{dE_{bond}}{dA} = 2\gamma_s$$

The value of G at the point of crack growth instability, referred to as Griffith's critical energy release rate, G_c, would then be

$$G_c = -\frac{dU}{dA} = \frac{dE_{bond}}{dA}$$

$$G_c = 2\gamma_s \tag{3.6}$$

We may therefore replace the $2\gamma_s$ term in Eqn. (3.3) with the Griffith critical strain energy release rate, G_c. This energy rate is rarely observed in practice and should be considered as a lower bound on the brittle fracture of materials. Note that G_c is a function of crack area not time, as in the traditional definition of rate. G_c may also replace the $2(\gamma_s + \gamma_p)$ in Eqn. (3.4) as the strain energy released at the critical point is the only source of energy available for *any* energy sinks associated with crack growth (such as plastic work at the crack tip). The fracture stress can now be written as

$$\sigma_f = \sqrt{\frac{G_c E}{\pi a}} \tag{3.7}$$

Equation (3.7) is often referred to as *Griffith's equation*. The critical energy release rate may be determined experimentally. This will be discussed later in the section on experimental methods.

Concept Challenge 3.2

Griffith's equation is applicable under limited conditions. What is the relevant plate geometry? How must the crack be oriented? What type of loading is permissible?

Example 3.1

Two geometrically equivalent rectangular panels, one made of glass the other made of polystyrene, are subjected to equal tensile loads. If a ½ inch through crack is present in both, which material would fail first when subjected to an increasing tensile load? The modulus of glass is 70,000 MPa and the critical energy release rate is about 7 J/m². The modulus of polystyrene is 3,000 MPa and the critical energy release rate is about 40 J/m².

We can convert MPa to Pa (1 MPa = 10⁶ Pa) and inches to meters (1 in = 0.0254 m).

For glass, we have

$$\sigma_{f_{glass}} = \sqrt{\frac{7 \times 70000 \times 10^6}{\pi\left(12.7 \times 10^{-2}\right)}} = 1.11 \times 10^6 \, \text{N/m}^2$$

For polystyrene, we have

$$\sigma_{f_{poly}} = \sqrt{\frac{40 \times 3000 \times 10^6}{\pi\left(12.7 \times 10^{-2}\right)}} = 0.55 \times 10^6 \, \text{N/m}^2$$

Alternately, since the crack size is the same for the glass and polystyrene, i.e. constant, we see from Eqn. (3.7) that

$$\sigma_f \propto \sqrt{G_c E}$$

We could therefore write

$$\frac{\sigma_{f_{glass}}}{\sigma_{f_{poly}}} = \sqrt{\frac{7 \times 70000}{40 \times 3000}} = 2.02$$

$$\sigma_{f_{glass}} = 2.02 \, \sigma_{f_{poly}}$$

The failure stress with a ½ inch crack in a glass panel is more than twice the failure stress with a ½ inch crack in a polystyrene panel. The polystyrene panel would therefore fail before the glass panel does.

Example 3.2

A thin sheet (5 mm) of magnesium oxide ceramic has dimensions 1 m x 0.6 m. The sheet has a small central crack whose axis is parallel to the 0.6 m sides. If the sheet is subjected to a tensile load of 21.14 kN normal to one of the 0.6 m sides with the opposite end fixed, what length does the crack have to be for the ceramic sheet to fracture? Assume the following material property values: Young's modulus E = 300 GPa; the surface energy is given as 2.6 J/m^2.

The applied stress (normal) to the crack axis is given by

$$\sigma = \frac{21.14 \times 10^3}{0.6 \times 5.0 \times 10^{-3}} = 7.05 \, \text{MPa}$$

We can now apply Eqn. (3.3)

$$7.05 \times 10^6 = \sqrt{\frac{2 \times 2.6 \times 300 \times 10^9}{\pi a}}$$

$$a = 0.01 \, \text{m}$$

The crack length for fracture is therefore 2a = 2 cm.

Concept Challenge 3.3

Could this approach be used if the sheet were subjected to tension and bending loads? Why? Why not?

3.3   DRIVING FORCE AND RESISTANCE TO CRACK GROWTH

Griffith's model is based on global energy balance, and states that fracture will occur if the energy stored within the structure E (mechanical strain energy + interatomic bond energy) exceeds the surface energy of the material. The approach assumes the material is brittle. For engineering materials that exhibit some ductility, Irwin [2] and Orowan [3] modified the energy release rate G to incorporates plastic strain as given in Equation (3.7). The model proposed that the majority of strain energy released was not used to create a new surface but dissipated as a result of plastic deformation near the crack tip as shown in Figure 3.2. It should be noted that the plastic deformation is small and confined to a thin layer around the crack tip.

Consider a thin infinite plate that contains a central crack of length 2a, subjected to a remote tensile stress σ, as shown in Figure 3.3. The energy release rate is given by

$$G = \frac{\pi \sigma^2 a}{E} \tag{3.8}$$

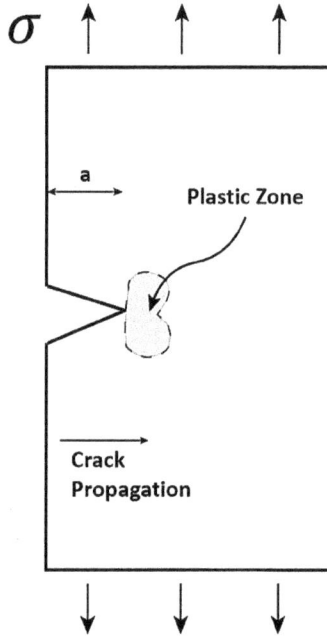

FIGURE 3.2 Plastic zone around crack tip.

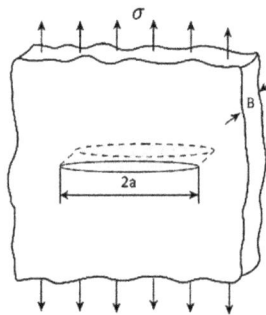

FIGURE 3.3 Large plate containing a through-thickness crack subjected to a remote, uniform tensile stress.

If we combine Equation (2.12), based on the stress intensity criterion, and the energy release rate theory of failure, Equation (3.8), by substituting for σ, we have

$$G = \frac{K_I^2}{E} \tag{3.9}$$

This is one of the most useful equations in linear fracture mechanics, and G represents the material's resistance to crack extension and has also been referred to as the crack driving force. Note that this relationship was derived based on a thin infinite plate with a central crack subjected to tension, and therefore in a plane stress state. If a plane strain condition exists, it can be shown that

$$G = \frac{K_I^2\left(1-v^2\right)}{E} \tag{3.10}$$

Irwin, however, used crack closure analysis to show that Equations (2.13–2.14) were true for all configurations. If there are mixed modes, the relationship between G and K_I becomes

$$G = \frac{K_I^2}{E'} + \frac{K_{II}^2}{E'} + \frac{K_{III}^2}{2\mu} \tag{3.11}$$

where $E' = E$ for plane stress or $\dfrac{E}{1-v^2}$ for plane strain, and μ is the shear modulus of the material.

Crack extension occurs when $G \geq G_{IC}$. The term G_{IC} incorporates all sources contributing to energy dissipation, including plastic flow which is a measure of the material's resistance to crack growth or fracture toughness. Table 3.1 gives the GIC and KIC values for some common materials.

Concept Challenge 3.4

According to the Griffith criterion, when does unstable crack growth occur in a material?

Example 3.3

A 3 mm thick 2024 T3 aluminum alloy plate contains a central crack 15 mm long. The plate is 185 mm wide and has a modulus of elasticity of 73.8 GPa, a tensile yield strength of 325 MPa, and a fracture toughness of 36 MPa√m. Determine the critical (a) failure stress and (b) strain energy release rate.

Does the material satisfy the necessary requirement for using the energy approach? Yes, because the plate is a high strength aluminum alloy, and is reasonably thick, the plastic deformation will be small enough compared to the rest of the material which will deform in a linear elastic fashion.

At fracture, the stress intensity factor is given by

$$K_{IC} = \alpha\sigma_f\sqrt{\pi a}$$

Is the plate sufficiently "large" to be considered infinite? Can we assume $\alpha = 1$? Let's check!

$$\frac{2a}{W} = \frac{15\,\text{mm}}{185\,\text{mm}} = 12.33$$

TABLE 3.1
Fracture Toughness for Various Materials [4]

Material	G_{IC}[kJ/m²]	$K_{IC}\left[\text{MPa}\sqrt{\text{m}}\right]$
Steel alloy	107	150
Aluminum alloy	20	37
Wood	6	8
Polystyrene	0.4	1.1
Refractory porcelain	0.020	1.0
Soda-lime glass	0.007	0.7

Generally, if $\dfrac{2a}{W} \geq 10$, we can assume the plate is large enough to ignore edge effects and assume $\alpha = 1$.

(a) The critical stress σ_c,

$$\sigma_c = \frac{K_{IC}}{\sqrt{\pi a}} = \frac{36\,\text{MPa}\sqrt{\text{m}}}{\sqrt{\pi\,(0.0075)}} \approx 234.5\,\text{MPa}$$

(b) The critical strain energy release rate is

$$G_{IC} = \frac{K_{IC}^2}{E} = \frac{\left(36\,\text{MPa}\sqrt{\text{m}}\right)^2}{73.8\times10^3\,\text{MPa}} = 17.56\,\text{kPa}$$

Concept Challenge 3.5

Can the modified Griffith theory be applied to a thin sheet of ductile low carbon steel containing a penny-shaped crack?

3.4   R-CURVE BEHAVIOR

So far, we have defined fracture and subsequent failure as unstable crack growth when the material's resistance to fracture is reached due to external loading. However, in thin sections, there is usually a period of stable crack propagation prior to unstable failure. Recall that, for a brittle material based on Griffith's energy criterion, crack growth will occur when $G = 2\gamma_s$.

However, this does not tell us anything about the stability of the resulting crack extension. If we let $R = 2\gamma_s$, where R represents the material resistance to crack growth, Equation (3.6) can be expressed as $G = R$. Irwin [5] modified the Griffith energy criterion proposing that, under plane stress conditions, stable crack growth will occur when

$$\frac{dG}{da} = \frac{dR}{da}$$

(3.12)

and unstable crack growth when

$$\frac{dG}{da} > \frac{dR}{da}$$

(3.13)

Crack stability depends on the variation of G and γ_s with crack size. If the quantity R is plotted against the crack extension, the resulting curve is referred to as a resistance curve or R-curve as shown in Figure 3.4. In this case, a plane stress section contains an initial crack of length a_0, subjected to a remotely applied stress σ_1. As the stress is increased to σ_2, there is a very small change in crack length as the material's resistance to crack growth also increases. If the stress continues to increase, stable crack growth will continue to occur until point I is reached on the curve, which corresponds to the instability point associated with the value G_C, beyond which unstable crack propagation will occur for increasing stress. The stable crack growth can be explained by the development of a plane strain section that is confined to the center

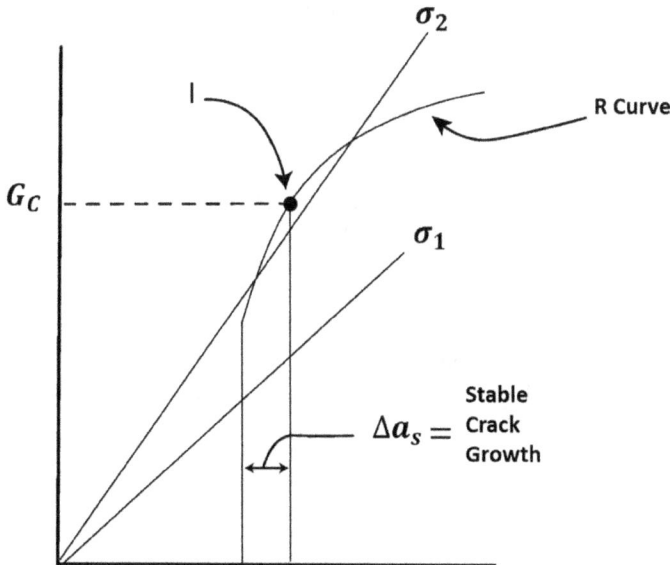

FIGURE 3.4 Material exhibiting R-curve behavior.

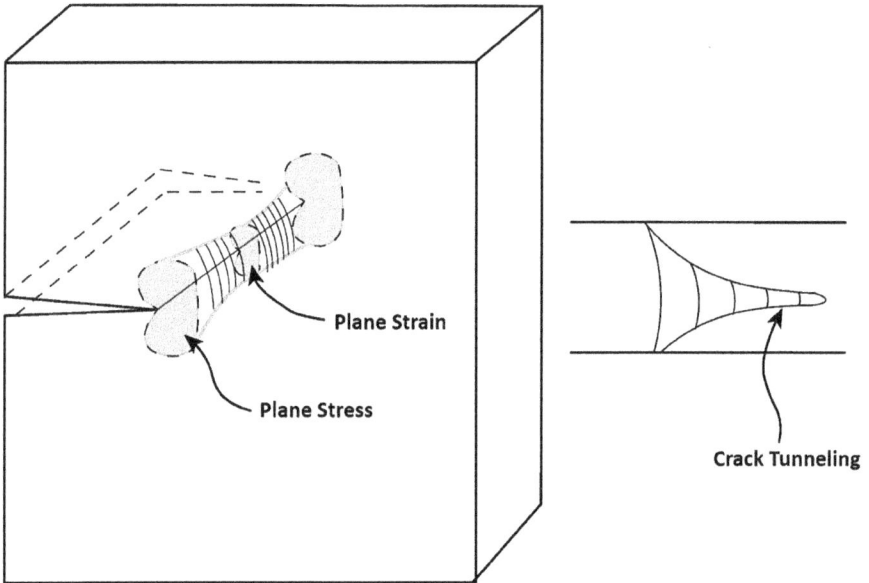

FIGURE 3.5 Crack tunneling in thin specimens.

of the section. The outer surface undergoes larger plastic deformation resulting in slower but stable crack growth. As the loading continues to increase the center of the crack advances first while the surface lags giving rise to the tunneling effect depicted in Figure 3.5. The R-curve behavior is independent of specimen geometry [6] and can thus be treated as a material property for a given material and thickness.

3.5 STRAIN ENERGY DENSITY

In Chapter 1, the concept of strain energy was introduced as the internal work done in deforming a body when subjected to externally applied forces. The action of creating strain in a material results in strain energy being stored in it. When the bar of length L, shown in Figure 3.6(a), is subjected to slow but steadily increasing force F, the external work done in creating deformation δ_x is stored internally as strain energy. As long as the bar is not displaced beyond its elastic limit, it will return to a zero strain energy state once the load is removed.

The shaded area under the force–displacement curve shown in Figure 3.6(b) is a measure of the work done, W, and is given by

$$dW = \frac{1}{2} F \delta_x$$

(3.14)

Now consider an elemental cube within the bar with unidirectional normal stress σ_{xx} arising in the x-direction from the externally applied force F, as shown in Figure 3.7(a).

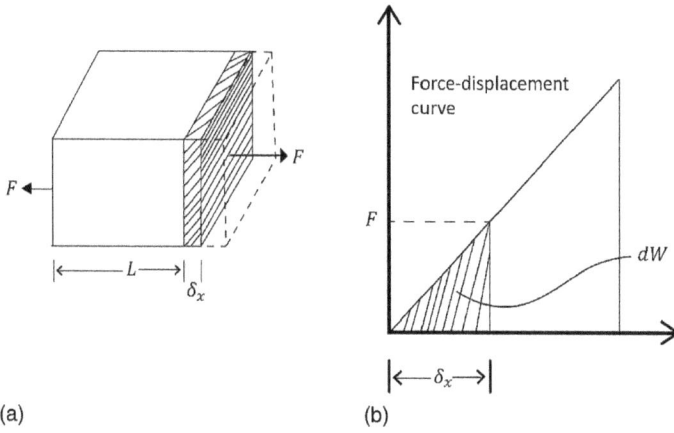

FIGURE 3.6   Bar subjected to (a) steadily increasing force, (b) Force-displacement curve.

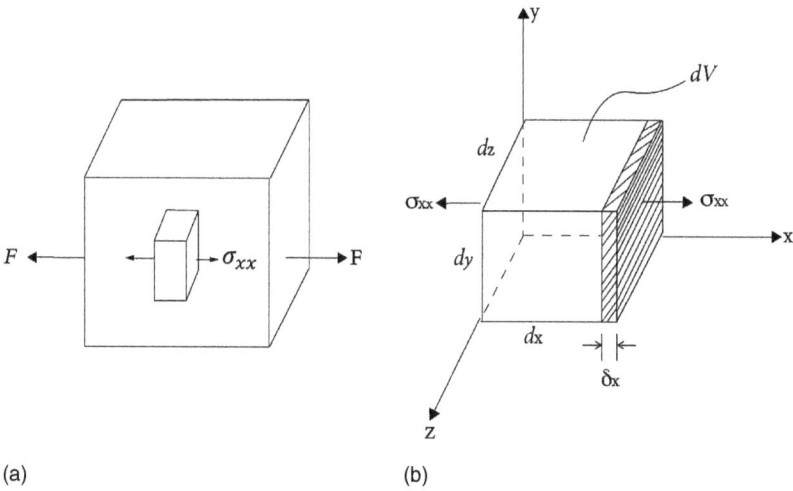

FIGURE 3.7   Normal stress on a differential volume element.

Since the volume of the cube, dV, is $dxdydz$ (see Figure 3.7(b)) and the resulting deformation in the x-direction is given by $\delta_x = \epsilon_{xx}dx$, we can express the work done on the volume element as

$$W = \frac{1}{2}\sigma_{xx}dydz \cdot \epsilon_{xx}dx \tag{3.15}$$

The strain energy per unit volume u, also known as the strain energy density, can now be defined as

$$u = \frac{W}{V} = \frac{1}{2}\sigma_{xx}\epsilon_{xx} \tag{3.16}$$

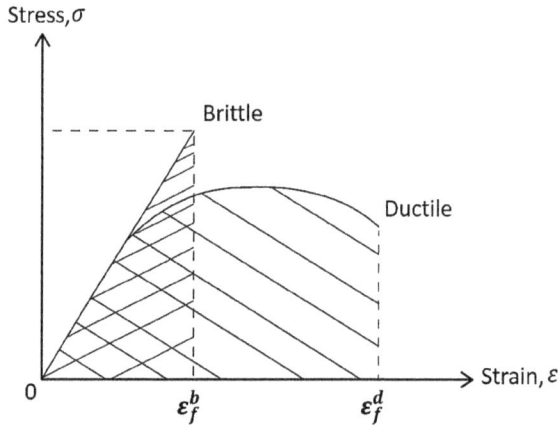

FIGURE 3.8 Stress-strain curves for brittle and ductile materials (shaded region represents strain energy density u).

If we examine the stress–strain curve shown in Figure 3.8, the shaded region is the area under the curve and represents the strain energy density u, which is also a measure of the fracture toughness of the material. It is useful to note the significantly larger area and thus fracture toughness for the ductile material shown, compared to the brittle under similar loading. The strain energy density has units of J/m³, and represents the energy required to deform the material. The total strain energy in the bar can be expressed as

$$U = \int_0^\varepsilon \sigma d\varepsilon$$

(3.17)

Making use of Hooke's law ($\sigma = E\varepsilon$), the elastic strain energy can be expressed as

$$U = \int_0^{\varepsilon_y} E\varepsilon d\varepsilon = \frac{1}{2} E\varepsilon_y^2 = \frac{\sigma_y^2}{2E}$$

(3.18)

where σ_y and ε_y represent the stress and corresponding strain at yield.

Example 3.4

A circular rod of length 2.5 m has an applied load of 2 kN in the axial direction. Given that the rod has a radius of 5 mm and Young's modulus of 180 GPa, determine the resulting strain energy in the rod.

Making use of Hooke's law for elastic materials ($\sigma = E\epsilon$) and substituting for $\sigma = F/A$ in Equation (3.16) we can write

$$W = \frac{F^2 dV}{2A^2 E}$$

Since the cross-sectional area can be assumed to be constant, W can be expressed as

$$W = \frac{F^2 dx}{2AE}$$

where dx is the length of the small element within the bar. The total strain energy in the bar is

$$U = \int_0^L \frac{F^2 dx}{2AE} = \frac{\left(2\times10^3\,\text{N}\right)^2 (2.5\text{m})}{2\pi\left(5\times10^{-3}\,\text{m}\right)^2 \left(1.8\times10^{11}\,\text{Pa}\right)}$$

$$= 0.354\,\text{J}$$

TRACTIONS

Consider a body in equilibrium with external forces F_1 and F_2 with a cut taken through the entire solid as shown in Figure 3.9(a). In order to maintain equilibrium after sectioning (Figure 3.9(b)), equal and opposite forces must be present on both sectioned surfaces. Point P, which is coincident to both surfaces, has an internal force **T** acting over the cross-section's area. A force of equal magnitude but in the opposite direction (−**T**) is required to maintain equilibrium. **T** is referred to as a traction vector and is given by

$$\mathbf{T} = \frac{F}{A} \qquad\qquad (3.19)$$

where A is the area of the shaded surface.

Although **T** has the same units as stress, MPa, it is a force vector acting over an area and not a stress tensor. The normal stress σ_n and shear stress τ_n are the components of the traction vector **T**, oriented normal and parallel to surface A, respectively. If **n** and **s** are unit vectors normal and parallel to the angled surface

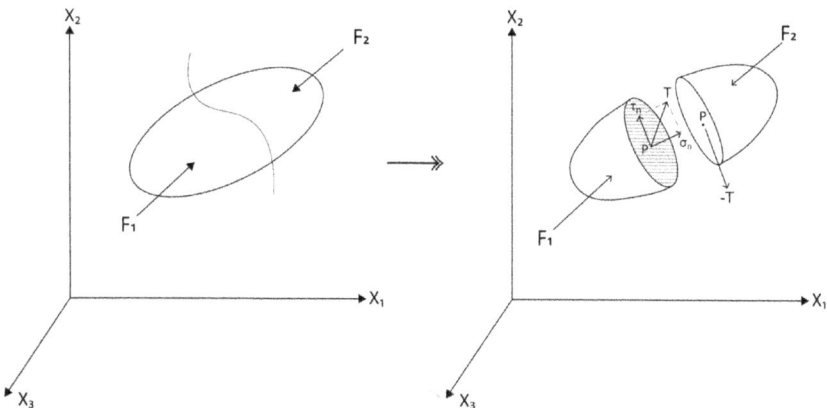

FIGURE 3.9 Traction vector at point P on sectioned body.

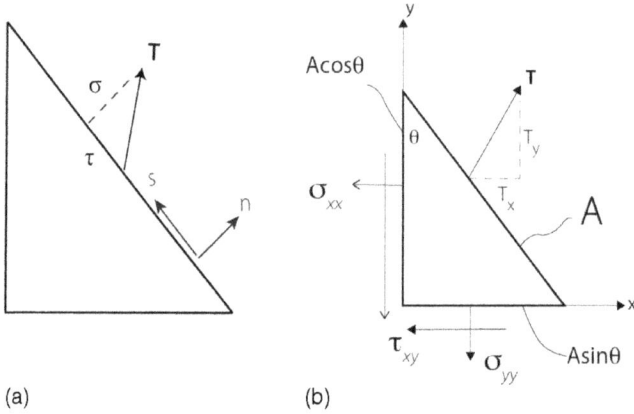

FIGURE 3.10   (a) Normal and shear stress components of traction vector; (b) Traction vector components on plane stress element.

shown in Figure 3.10(a), the components of **T** in the **n** and **s** directions can be obtained from

$$\sigma = \mathbf{T} \cdot \mathbf{n} \text{ and } \tau = \mathbf{T} \cdot \mathbf{s} \tag{3.20}$$

The traction vector and state of stress at the point shown in Figure 3.10(b) can be related by imposing equilibrium conditions on the element which leads to

$$\sigma_{xx} A\cos\theta + \tau_{xy} A\sin\theta = T_x A \tag{3.21a}$$

$$\tau_{xy} A\cos\theta + \sigma_{yy} A\sin\theta = T_y A \tag{3.21b}$$

Since A is common to both sides of each equation, this can be further reduced to

$$\sigma_{xx} \cos\theta + \tau_{xy} \sin\theta = T_x \tag{3.22a}$$

$$\tau_{xy} \cos\theta + \sigma_{yy} \sin\theta = T_y \tag{3.22b}$$

The unit normal to the surface shown in Figure 3.9(a) can be expressed as

$$\mathbf{n} = \left(\cos\theta, \sin\theta\right) = \left(n_x, n_y\right) \tag{3.23}$$

Substituting for n_x and n_y in Equation 3.22 gives

$$\sigma_{xx} n_x + \tau_{xy} n_y = T_x \tag{3.24a}$$

$$\tau_{xy} n_x + \sigma_{yy} n_y = T_y \tag{3.24b}$$

These can now be written as

$$\mathbf{T} = \sigma \cdot \mathbf{n} \tag{3.25}$$

Example 3.5

A uniform bar having a cross sectional area of 450 mm² is subjected to a tensile force of 3,500 N in the x-direction. Determine the normal and shear stresses on a surface inclined at $\theta = 25^0$ as shown in Figure 3.11.
 Expressing the force in vector form

$$F = 3500 \mathbf{i}\, N$$

and the surface area of the bar A along the 25^0 inclined plane is

$$A = \frac{450\,mm^2}{\cos(25^0)} = 496.52\,mm^2$$

The traction vector is

$$T = \frac{F}{A} = \frac{3500 \mathbf{i}\, N}{496.52\,mm^2} = 7.05 \mathbf{i}\, MPa$$

The unit normal to the inclined surface is given by

$$n = \left(\cos 25^0, \sin 25^0, 0\right)$$

The normal stress can now be determined from

$$\sigma = T \cdot n = (7.05,0,0) \cdot \left(\cos 25^0, \sin 25^0, 0\right) = 6.39\,MPa$$

The parallel surface vector is

$$s = \left(-\sin 25^0, \cos 25^0, 0\right)$$

The shear stress on the inclined surface is

$$\tau = T \cdot s = (7.05,0,0) \cdot \left(-\sin 25^0, \cos 25^0, 0\right) = -2.98\,MPa$$

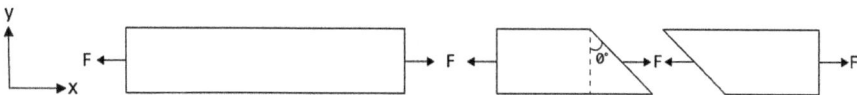

FIGURE 3.11   Uniform bar in tension.

3.6 THE J-INTEGRAL

In cases where plastic deformation around the crack tip becomes large ($r \geq a$), and quantities such as K and G do not adequately characterize the elastic-plastic behavior near the crack tip, an alternate approach known as the J-integral can be employed to describe crack extension in nonlinear elastic materials.

The J-integral concept was formulated by Rice [7] and defines a line integral that is path independent, having a value which is equal to the decrease in potential energy when the crack advances by an increment of crack growth da. This is similar to the strain energy release rate G, when dealing with a linear elastic body. In fact, in such a situation J and G would be equivalent, making it unnecessary to use the more complex J-integral formulation. The usefulness of J is in elastic plastic fracture mechanics (EPFM), where the body is considered nonlinear elastic. It should be noted that application of the J-integral is restricted to nonlinear elastic materials that do not undergo unloading in any part of the body. Figure 3.12 illustrates the differences in unloading behavior for linear and nonlinear bodies undergoing plastic deformation. For ASTM A335B steel under large scale yielding, the plastic zone is estimated to be around 3.5 mm in size.

Consider a two-dimensional crack in a tough material subjected to large scale yielding as shown in Figure 3.13. We can define an arbitrary counter-clockwise contour Γ, starting at point O and traveling along the crack surface, eventually extending outward into the surrounding material such that it encapsulates the crack tip.

The value of J can be determined from

$$J = \int_{\Gamma} \left(U dy - T_i \frac{\partial u_i}{\partial x} \right) ds$$

(3.26)

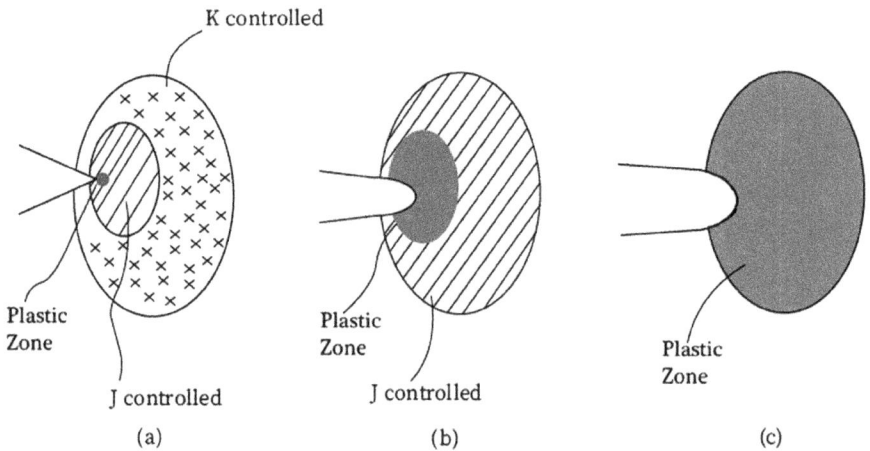

FIGURE 3.12   (a) Small scale yielding; (b) Elastic-plastic behavior; (c) Large scale yielding.

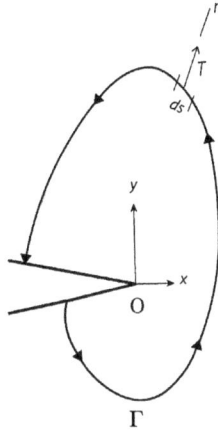

FIGURE 3.13 J integral defined by closed contour Γ.

where
    J = effective energy release rate (MPa · m)
    U = strain energy density (J/m³)
    T_i = components of the traction vector
    u_i = displacement vector components
    ds = length increment along contour Γ

The strain energy density is defined as

$$U = \int_0^{\varepsilon_{xx}} \sigma_{xx} d\varepsilon_{xx} + \int_0^{\varepsilon_{xy}} \sigma_{xy} d\varepsilon_{xy} + \int_0^{\varepsilon_{xz}} \sigma_{xz} d\varepsilon_{xz} + \cdots \tag{3.27}$$

where σ_{xy} and σ_{xz} represent shear stresses τ_{xy} and τ_{xz} respectively. The elastic stresses can be expressed in matrix form as

$$\sigma_{ij} = \begin{bmatrix} \sigma_{xx} & \tau_{xy} & \tau_{xz} \\ \tau_{yx} & \sigma_{yy} & \tau_{yz} \\ \tau_{zx} & \tau_{zy} & \sigma_{zz} \end{bmatrix} \tag{3.28}$$

and the traction vector at a given point on the contour is given by

$$T = T_x + T_y + T_z \tag{3.29}$$

where T_x, T_y, and T_z are defined as in Equation (3.24a). It should be noted that application of the J-integral is restricted to nonlinear elastic materials that do not undergo unloading in any part of the body.

Concept Challenge 3.6

Using the small strain relationship $\varepsilon_{xx} = \dfrac{\partial u_x}{\partial x}$, determine the strain if the displacement of the body along the x-direction is given by $u_x = 3x - y$.

Example 3.6

Determine the line integral $\oint \vec{T}.\dfrac{\partial \vec{u}}{\partial x} ds$ around the closed curve shown in Figure 3.14, where $\vec{T} = \vec{i} + \vec{j}$ and $\dfrac{\partial \vec{u}}{\partial x} = 3x\vec{i} + 3y\vec{j}$.

The dot product $\vec{T}.\dfrac{\partial \vec{u}}{\partial x}$ yields $(1)(3x) + (2)(3y) = 3x + 6y$, therefore

$$\oint 3x + 6y\, ds =$$

$$= \int_1^4 \left(3x+6y\right)_{y=1} dx + \int_1^3 \left(3x+6y\right)_{x=4} dy + \int_4^1 \left(3x+6y\right)_{y=3}\left(-dx\right)$$

$$+ \int_1^3 \left(3x+6y\right)_{x=1}\left(-dy\right)$$

$$= \int_1^4 3x+6\, dx + \int_1^3 12+6y\, dy + \int_4^1 -3x-18\, dx + \int_3^1 -3-6y\, dy$$

$$= 40.5 + 48 + 76.5 + 10 = \boxed{175}$$

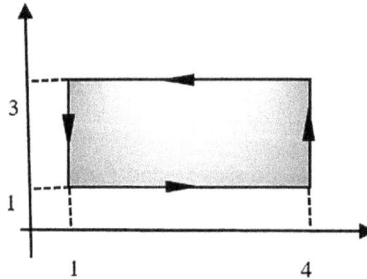

FIGURE 3.14   Line Integral around a Rectangular Path.

Example 3.7

Determine the line integral $\oint \vec{T}.\dfrac{\partial \vec{u}}{\partial x} ds$ around the closed curve shown in Figure 3.15 where the traction vector, $\vec{T}$, is given by $\vec{T} = \vec{i} + \vec{j}$ and $\dfrac{\partial \vec{u}}{\partial x} = 3x\vec{i} + 3y\vec{j}$.

The line integral

$$\oint \vec{T}.\dfrac{\partial \vec{u}}{\partial x} ds = \oint 3x + 6y\, ds$$

$$= \int_a 3x+6y\, ds + \int_b 3x+6y\, ds + \int_c 3x+6y\, ds$$

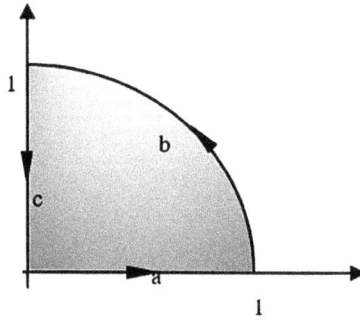

FIGURE 3.15 Line Integral around a Quarter Circular Path.

For the quarter circular path AB, the curve can be parameterized by $x = \cos t$, $y = \sin t$ where $0 < t < \dfrac{\pi}{2}$. The elemental length ds along the curve can be expressed as

$$ds = \sqrt{(dx)^2 + (dy)^2}$$
$$= dt\sqrt{\left(\frac{dx}{dt}\right)^2 + \left(\frac{dy}{dt}\right)^2}$$

In this case,

$$\frac{dx}{dt} = -\sin t, \frac{dy}{dt} = \cos t$$

$\int_{AB} 3x + 6y\, ds$ therefore becomes

$$\int_{t=0}^{t=\frac{\pi}{2}} (3\cos t + 2\sin t)\sqrt{(-\sin t)^2 + (\cos t)^2}\, dt$$
$$= \int_{t=0}^{t=\frac{\pi}{2}} 3\cos t + 2\sin t\, dt$$
$$= \left|3\sin t - 2\cos t\right|_0^{\frac{\pi}{2}}$$
$$= 5$$

For BC $ds = -dy$

$$\int_{BC} 3x + 6y\, ds = \int_1^0 \left|3x + 6y\right|_{x=0} (-dy)$$
$$= \left|-3y^2\right|_1^0$$
$$= 3$$

For CA $ds = dx$

$$\int_{CA} 3x + 6y\, ds = \int_{1}^{0} |3x + 6y|_{y=0} \, (dx)$$

$$= \left| \frac{3x^2}{2} \right|_{0}^{1}$$

$$= \frac{3}{2}$$

$$\oint \vec{T}.\frac{\partial \vec{u}}{\partial x}\, ds = 5 + 3 + \frac{3}{2} = \boxed{\frac{19}{2}}$$

Example 3.8

The double cantilever beam of length L and depth b has a shear load applied to the surfaces of the free end as shown in Figure 3.16. Assume plane strain conditions, and that the crack surface is unloaded. Determine the value of J.

In order to evaluate J, a line integral path must be selected that encloses the crack tip such that the initial and final end points lie on the two crack surfaces, as shown. Note the line is actually coincident with the boundary of the cantilever but is offset in the figure for illustration. Recall the J-integral is given by

$$J = \int_{\Gamma} \left(U dy - T_i \frac{\partial u_i}{\partial x} \right) ds$$

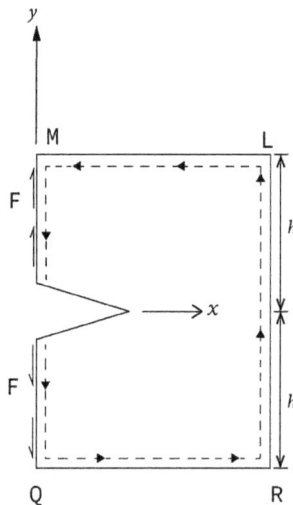

FIGURE 3.16   Double cantilever beam subjected to shear loading.

The integral can be evaluated as:

$$J = \int_{PQ} + \int_{QR} + \int_{RL} + \int_{LM} + \int_{MN}$$

Since contours QR, RL, and LM are free surfaces, the tractions are zero. In addition, the crack is considered to be a line (exaggerated in Figure 3.16 for illustration only), meaning $dy = 0$. The strain energy density term U along the line path is also very small and can also be neglected. This leaves only segments PQ and MN for evaluation; the integral reduces to

$$J = \int_{PQ} + \int_{MN}$$

$$J = \int_{PQ}\left(0 - \left(T_x\frac{\partial u_x}{\partial x} + T_y\frac{\partial u_y}{\partial x}\right)\right)ds + \int_{MN}\left(0 - \left(T_x\frac{\partial u_x}{\partial x} + T_y\frac{\partial u_y}{\partial x}\right)\right)ds$$

The traction vector component $T_x = 0$ on surfaces PQ and MN. Using limits of integration ranging from 0 to h we have

$$J = \int_0^h\left(-T_y\frac{\partial u_y}{\partial x}\right)ds + \int_0^h\left(-T_y\frac{\partial u_y}{\partial x}\right)ds$$

$$= \int_0^h\left(-T_y\frac{\partial u_y}{\partial x}\right)ds + \int_0^h\left(-T_y\frac{\partial u_y}{\partial x}\right)ds$$

$$= -2\int_0^h\left(T_y\frac{\partial u_y}{\partial x}\right)ds$$

In the case of the cantilever beam, the term $\dfrac{\partial u_y}{\partial x}$ is the slope at the free end given by $\dfrac{FL^2}{2EI}$ Substituting in J gives

$$J = -2\int_0^h\left(T_y\frac{FL^2}{EI}\right)ds = \boxed{-\frac{12FL^2}{Ebh^3}\int_0^h T_y ds}$$

In practical cases, where stresses and strains can occur in the thickness direction, the evaluation of the J-integral becomes more complex and the use of finite element techniques are normally necessary. An actual example where the J-integral can be used for describing crack extension is discussed in Section 3.7.

3.7 CRACK TIP OPENING DISPLACEMENT (CTOD)

Another popular approach when dealing with extensive yielding in the crack-tip region is the crack tip opening displacement (CTOD). The concept was first introduced by Wells [8] when attempting to determine the fracture toughness of medium

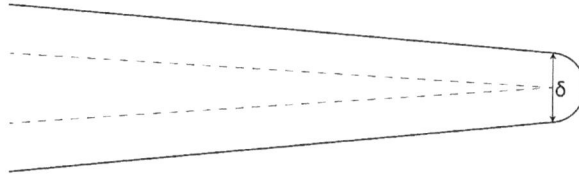

FIGURE 3.17   Blunted crack.

strength structural steel. He discovered that it could not be determined by linear elastic fracture mechanics (LEFM). Many materials used in the design and fabrication of engineering components fall into this category, including large structures such as water tanks and pressure vessels. In contrast to the stress intensity method described in Chapter 2, CTOD uses strains within the crack-tip region rather than stresses to estimate the material's resistance to fracture.

Wells noticed that in fractured specimens prior to fracture, plastic deformation had blunted the sharp crack tip that was initially present as shown in Figure 3.17. He proposed that the displacement δ, at the crack tip be used as a measure of fracture toughness for materials experiencing elastic-plastic or fully plastic behavior prior to failure.

Later on, Burdekin and Stone [9] proposed a strip yield model, as depicted in Figure 3.18, to estimate the CTOD δ, in an infinite plate containing a through crack of length $2a$ given by

$$\delta = \frac{8a\sigma_{ys}}{\pi E} \ln\left[\sec\left(\frac{\pi}{2} \frac{\sigma}{\sigma_{ys}} \right) \right] \tag{3.30}$$

where σ is a uniform applied stress. In the case of small scale yielding, CTOD can be related to K_I and G in the following way

$$\delta = \frac{K_I^2}{\sigma_{ys}E} = \frac{G}{\sigma_{ys}} \tag{3.31}$$

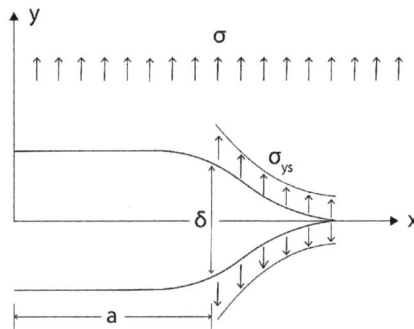

FIGURE 3.18   Strip yield model proposed by Burdenkin and Stone for estimating CTOD.

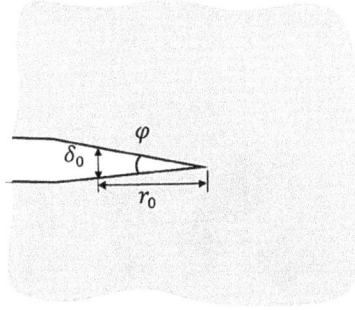

FIGURE 3.19 Crack Tip Opening Angle (CTOA) determined for a constant crack opening displacement δ_0 at a distance r_0 from the crack tip.

While the strip yield model is based on plane stress conditions and assumes no strain hardening occurs, the relationship between CTOD and K_I and G does depend on the state of stress and strain hardening.

The crack tip opening angle (CTOA) proposed by Rice [10] is defined as the average angle φ, measured at a finite distance from the crack tip as shown in Figure 3.19.

The CTOA parameter was developed primarily to characterize stable crack extension behavior in thin-walled materials in low constraint conditions as used in aircraft engineering and the pipeline industry [11]. CTOA can be determined from

$$\text{CTOA} = 2\tan^{-1}\left(\frac{\delta_0}{2r_0}\right)$$

(3.32)

where δ_0 is the crack opening displacement at a distance r_0 behind the crack tip.

3.8 APPLICATIONS

CASE STUDY 1: PREDICTION OF STABLE CRACK GROWTH ON AIRCRAFT FUSELAGE

During the past three decades, standards have been introduced to enable reproducible measurements of CTOA including improved computer-aided photographic methods that have been developed to measure surface CTOA during the fracture process [12].

In the FAA and NASA aging aircraft program several analytical tools were developed to predict the residual strength of a damaged aircraft fuselage. This included prediction of failure of an actual aircraft fuselage that contained multiple site damage (MSD) cracking similar to the DC-9 aircraft shown in Figure 3.20 [13]. A finite element code STAGS (structural analysis of general shells) was used in conjunction with the CTOA fracture criterion. The CTOA for Al 2014-T3 sheets, the aft-bulkhead material, was needed to perform a

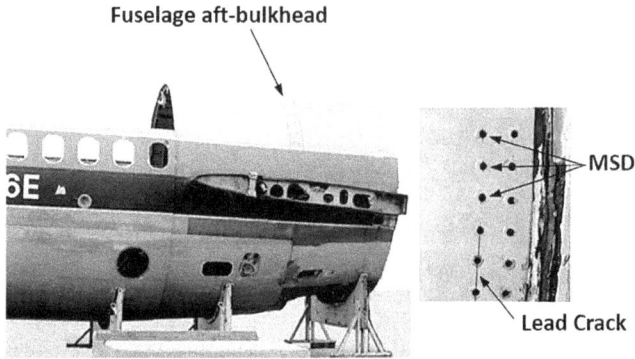

FIGURE 3.20   DC-9 fuselage aft-bulkhead showing multiple site damage (MSD) and lead crack locations.

stable tearing test using the STAGS code but was not available in the published literature. It was instead estimated using test data from the *Damage Tolerant Design Handbook*, MCIC-HR-01R for Al 2014-T6 sheets. A STAGS model was developed for the test specimens used in the handbook tests. The best match between the model and test data was found when using a CTOA = 3.4^0. A STAGS model of the aft-bulkhead was then developed as shown in Figure 3.21, for performing stable tearing analysis at various pressures.

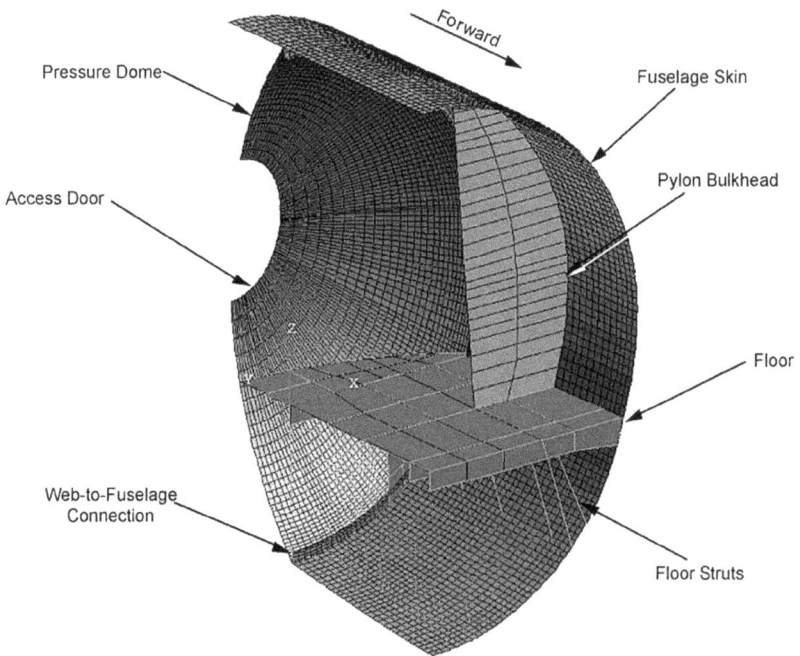

FIGURE 3.21   Overall view of STAGS model for aft pressure bulkhead analysis.

The model prediction indicated that the lead crack started to propagate at a cabin pressure level of approximately 56.5 kPa, linking up with the MSD in the first adjacent fastener hole at 64 kPa. The lead crack became unstable immediately after the first link-up. No indication of propagation for the MSD cracks was predicted. The predicted failure pressure correlated within 5% of the experimental failure pressure of 62 kPa.

CASE STUDY 2: FRACTURE ANALYSIS EXTERNAL THREAD ON A POSITIVE DISPLACEMENT MOTOR DRIVE SHAFT SHELL

The positive displacement motor (PDM) is a newer type of downhole motor that drives drill bits rotating to break rocks. Drilling with a PDM has many advantages including high rotation speed, small energy consumption, and easy operation [14]. The primary components of a PDM are shown in Figure 3.22. The failure mechanism of the threads has been previously studied using strength and fatigue analysis. However, there is limited research on the thread and the region containing the crack which is usually in an elastic-plastic state under working conditions. This study examines the elastic-plastic behavior of the external thread of the drive shaft shell (DSS) containing a circumferential semi-elliptical crack lying at the outer surface of the thread root.

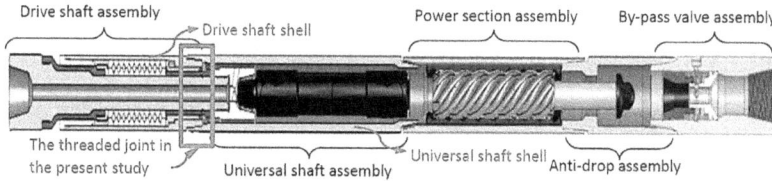

FIGURE 3.22 Components of a positive displacement motor.

The thread has a helix angle and the region near the crack is usually in an elastic-plastic state under working conditions. Due to the complexity of the problem, elastic fracture analysis is often conducted without considering the helix angle, while the error generated by this simplified model is rarely mentioned. The fracture performance of the cracked external thread of the DSS is simulated under the combined action of pre-load and bending moment. Effects of both plastic deformation and the helix angle on the fracture prediction are evaluated.

A three-dimensional finite element model containing a semi-elliptical surface crack on the root of the external joint was developed based on the sketch shown in Figure 3.23(a), using the FE finite element analysis software ANSYS. Figure 3.23(b) shows the crack surface, in which the shape of the crack is described by aspect ratio a/c and depth ratio a/t. Here a and c are the major and minor axes of the semi-ellipse, and t is the thickness of the cross-section where the crack is located. The deepest point of the crack front is denoted as point A, and the boundary point is B.

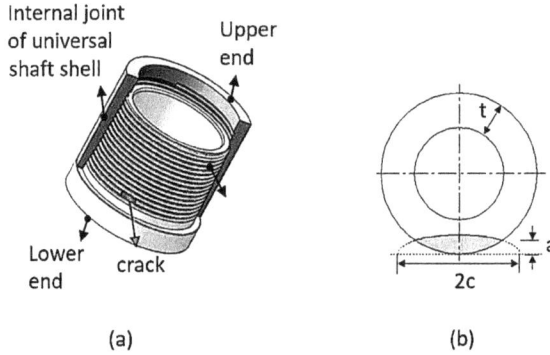

FIGURE 3.23 Sketch of DSS joint, (a) model of engaging thread, (b) cross-section containing crack.

Taking cracks $a/c = 0.4$ and $a/t = 0.1, 0.3, 0.5$, and 0.7 as examples, J-integrals are calculated based on an elastic and also an elastic-plastic model, respectively. The study revealed that the calculated J-integral based on the elastic-plastic constitutive relation is larger than the elastic result. The fracture evaluation is unsafe if the plastic deformation of the material is neglected in the model. The helix angle has influence on the predicted J-integral, and the numerical results are also dangerous if the helix angle is not taken into account.

PROBLEMS

3.1. A brittle material has a specific surface energy of 1.1 J/m and a modulus of elasticity of 70 GPa. It has an external crack with a length of 0.2 mm. What is the maximum amount of stress that will not cause crack propagation? What are some good questions to ask before attempting this problem?

3.2. The specific surface energy for a silica glass was determined as 3.6 J/m². Determine the largest sustainable crack for a critical stress value of 18 MPa given $E = 70$ GPa.

3.3. A large thin sheet of low alloy steel ($\sigma_Y = 1150$ MPa) used in the fabrication of a motor casing has a fracture toughness of 22 KJ/m². For safe operations the design requires a minimum factor of safety of 1.5. Determine the minimum size of the defect responsible for the brittle fracture of such a casing, if $E = 207$ GPa.

3.4. A large steel sheet containing a 40 mm central crack has a fracture stress of 320 MPa. Determine the critical strain energy release rate is if $E = 207$ GPa.

3.5. Calculate the fracture stress of a large steel sheet containing a central crack of length 90 mm, assuming plane strain conditions and given the following:

$$G_{IC} = 24\,\text{kJ/m}^2; E = 200\,\text{GPa}; v = 0.3$$

3.6. An aluminum bar of length 0.5 m and having a rectangular cross-section of 12×25 mm is subjected to an axial load of 480 kN. Determine the strain energy per unit volume if $E = 70$ GPa.

3.7. Calculate the traction vector on a surface with unit normal $\mathbf{n} =$ $(0.300, 0.500, 0.528)$, if the stress tensor is given by:

$$\sigma = \begin{bmatrix} 45 & 12 & 28 \\ 12 & 95 & 18 \\ 28 & 18 & 15 \end{bmatrix}$$

3.8. Determine the force acting on a surface of area 75 mm² with traction vector $\mathbf{T} = 43.62\,\mathbf{i} + 77.12\,\mathbf{j} + 33.29\,\mathbf{k}$ MPa.

3.9. A uniform bar having a cross-sectional area of 275 mm² is subjected to a tensile force of 2,500 N in the y-direction. Determine the normal and shear stresses on a surface inclined at $\theta = 22.5^0$ as shown in Figure 3.24.

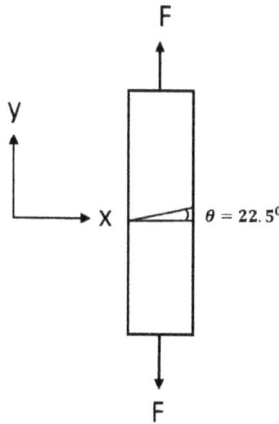

FIGURE 3.24

3.10. A large steel plate contains a through-the-thickness central crack with critical value of 8 mm and fracture stress of 278 MPa. Determine δ_c under small scale yielding conditions given

$$E = 200\,GPa;\ \sigma_{YS} = 1050\,MPa.$$

REFERENCES

[1] A. Griffith, "The Phenomenon of Rupture and Flow in Solids," *Philosophical Transactions of the Royal Society of London*, vol. 221, no. A, pp. 163–198, 1920.

[2] G. R. Irwin, "Fracture Dynamics. Fracturing of Metals," *American Society of Metals*, pp. 147–166, 1948.

[3] E. Orowan, "Fracture and Strength of Solids," *Reports on Proogress in Physics*, vol. XII, p. 185, 1948.

[4] J. E. Strawley and W. F. Brown, "Fracture Toughness Testing," *ASTM STP*, vol. 381, p. 133, 1965.

[5] G. R. Irwin, *ASTM Bulletin*, vol. 29, 1960.

[6] R. H. Heyer and D. E. McCabe, "Plane-Stress Fracture Toughness Testing using a Crack-Line-Loaded Specimen," *Engineering Fracture Mechanics*, vol. 4, pp. 393–412, 1972.

[7] J. R. Rice, "A Path Independent Integral and the Approximate Analysis of Strain Concentration by Notches and Cracks," *Journal of Applied Mechanics*, vol. 35, p. 379, 1968.

[8] A. A. Wells, "Unstable Crack Propagation in Metals: Cleavage and Fast Fracture," *Proceedings of the Crack Propagation Symposium*, vol. 1, no. 84, 1961.

[9] F. M. Burdekin and D. E. W. Stone, "The Crack Opening Displacement Approach to Fracture Mechanics in Yielding Materials.," *Journal of Strain Analysis*, vol. 1, pp. 144–153, 1966.

[10] J. R. Rice, "Mechanics and Mechanisms of Crack Growth," *British Steel Corporation Physical Metallurgy Centre Publication*, pp. 14–39, 1975.

[11] X. K. Zhu and J. A. Joyce, "Review of fracture toughness (G, K, J, CTOD, CTOA) testing," *Engineering Fracture Mechanics*, vol. 85, pp. 1–46, 2012.

[12] W. R. Tyson, J. C. Newman Jr and S. Xu, "Characterization of stable ductile crack propagation by CTOA: Review of theory and applications," *Fatigue & Fracture of Engineering Materials & Structures*, vol. 41, pp. 2421–2437, 2018.

[13] C. Hsu, J. Lo, J. Yu, X. Lee and P. Tan, "Residual strength analysis using CTOA criteria for fuselage structures containing multiple site damage," *Engineering Fracture Mechanics*, vol. 70, pp. 525–545, 2003.

[14] G. Zhao and G. Liao, "Elastic-plastic Fracture Analysis of External Thread of Drive Shaft Shell of a Positive Displacement Motor," *Mechanika*, vol. 26(5), pp. 375–382, 2020.

4 Applications

4.1 FATIGUE FAILURES

Fatigue is one of the most common types of failure in engineering structures accounting for almost 80% of all failures [1]. When materials experience large service stresses, the resulting defects can usually be detected and steps taken to prevent failure. However, with fatigue failures the applied stresses are typically below the elastic limit of the material.

Nevertheless, they can still result in catastrophic failure when small defects grow to critical lengths as previously discussed with static or monotonic loading. In this chapter we will consider crack growth in the presence of cyclic stresses and discuss some of the applications of fracture mechanics for predicting crack propagation in fatigue.

4.1.1 FATIGUE FUNDAMENTALS

Fatigue failure occurs in materials subject to fluctuating or cyclic loading. While fluctuating loads are not applied continuously compared with cyclic loading they do account for a wide application of cases including aircraft (Figure 4.1), bridges, and structural and machine components. Thermal effects such as those experienced in electrical components which are powered on and off can also produce cyclic stresses that result in fatigue failure of electrical joints. The applied stresses can be continuously constant or of variable amplitude as shown in Figure 4.2.

The common stress parameters are the mean stress (σ_m), alternating stress(σ_a), the stress ratio (R), and the stress amplitude (S_a) which are given by:

$$\sigma_m = \frac{\sigma_{max} + \sigma_{min}}{2} \tag{4.1a}$$

$$\sigma_a = \frac{\sigma_{max} - \sigma_{min}}{2} \tag{4.1b}$$

$$R = \frac{\sigma_{min}}{\sigma_{max}} \tag{4.1c}$$

$$S_a = \frac{\sigma_a}{\sigma_m} \tag{4.1d}$$

From a design standpoint, it would be useful to be able to determine the number of cycles N that a part can handle before unstable crack growth occurs.

DOI: 10.1201/9781003052050-4

Concept Challenge 4.1

An engineering component is subjected to cyclic loading. What parameters do you expect would influence the fatigue life of the part in terms of N?

FIGURE 4.1 Fatigue failure of landing gear in commercial aircraft.

Source: Image obtained from the National Transportation Safety Board, Aviation Investigation # DCA13FA131.

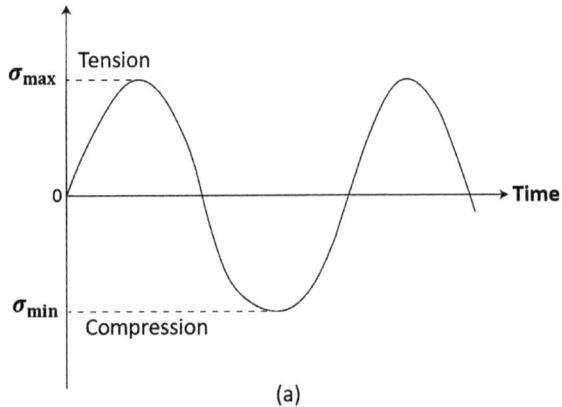

FIGURE 4.2 Stress cycles: (a) Tension-compression loading (symmetrical).

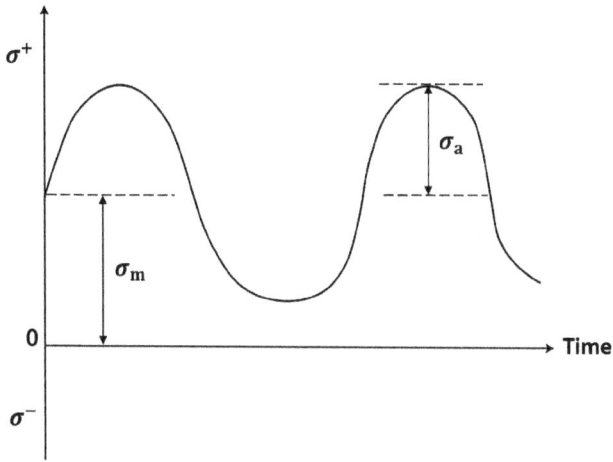

(b)

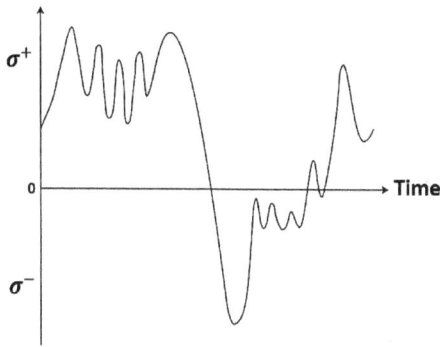

(c)

FIGURE 4.2 (CONTINUED) Stress cycles: (b) Tension-tension loading (symmetrical); (c) Random loading (asymmetric).

4.1.1.1 S-N Curves

In an effort to characterize the behavior of materials under fatigue loading the most common method is through a stress-cycle curve. Specimens are subjected to repeated or varying stresses and the cycles to failure counted. The data is represented in an *S-N* curve as shown in Figure 4.3.

Certain materials such as ferrous and titanium alloys when subject to cyclic loading, approach a stress value known as the fatigue or endurance limit S_e, below which the fatigue life is considered infinite. Nonferrous alloys such as aluminum, do not exhibit this type of behavior, instead the stress continues to decrease as the number

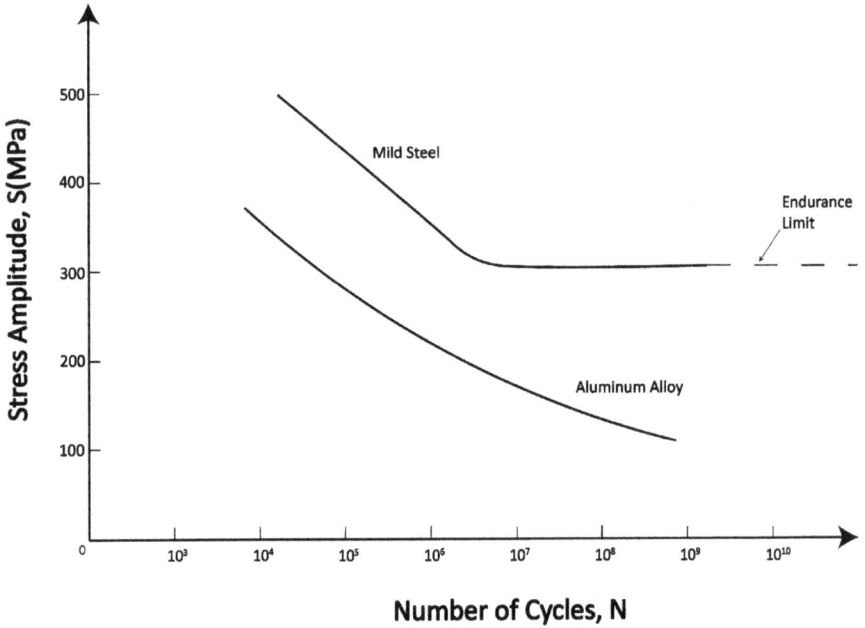

FIGURE 4.3   S-N curves for mild steel and aluminum alloy.

of cycles increases. In such cases, the fatigue life is characterized by the specific life (N) at a specified stress value known as the fatigue strength S_f.

Consider a material containing an initial crack subjected to a constant amplitude cyclic load, causing a plastic zone to develop at the crack tip. If the resulting zone is small and contained within the singularity dominated zone as shown in Figure 4.4, then linear elastic fracture mechanics (LEFM) can be applied.

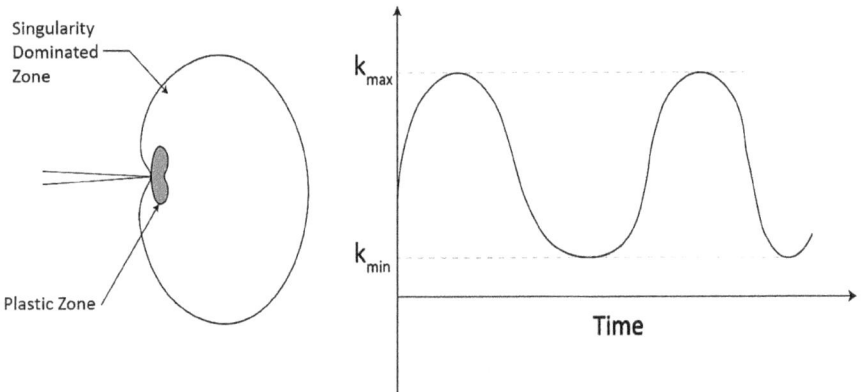

FIGURE 4.4   Fatigue crack subjected to constant stress amplitude under small-scale yielding.

This means the conditions around the crack tip at that instant can be characterized by the stress intensity factor K. It is convenient to represent the stress parameters in terms of K as

$$\Delta K = K_{max} - K_{min} \qquad (4.2)$$

$$R = \frac{K_{min}}{K_{max}} \qquad (4.3)$$

The crack growth per cycle, also known as the crack growth rate $\frac{da}{dN}$, can be expressed as a function of the applied stress intensity factor range ΔK and the stress ratio R as:

$$\frac{da}{dN} = f\left(\Delta K, R\right) \qquad (4.4)$$

Several functions have been proposed to define f, the majority of which are empirical. The fatigue life of a specimen can be determined by integrating Eqn. (4.4) which is given by

$$N = \int_{a_i}^{a_f} \frac{da}{f\left(\Delta K, R\right)} \qquad (4.5)$$

where a_i and a_f are the initial and final crack lengths respectively and N is the number of cycles required to grow the crack from a_i to a_f.

A plot of $\frac{da}{dN}$ against log ΔK as shown in Figure 4.5 illustrates three distinct regions typically associated with fatigue crack growth behavior in metals.

In Stage I, there is a stress intensity value ΔK_{th}, below which the crack growth rate is very low or does not occur at all. This threshold value allows us to calculate the maximum permissible crack length or the applied stress level to avoid the growth of a fatigue crack. If ΔK_{th} is exceeded, crack initiation occurs and the growth rate increases fairly rapidly as ΔK increases. Crack growth is dependent on factors such as microstructure, stress ratio, environmental conditions, and surface damage. In Stage II, the crack growth rate is considered linear and can be described using a power law expression which was first introduced by Paris and Erdogan [2] and given by:

$$\frac{da}{dN} = C\Delta K^n \qquad (4.6)$$

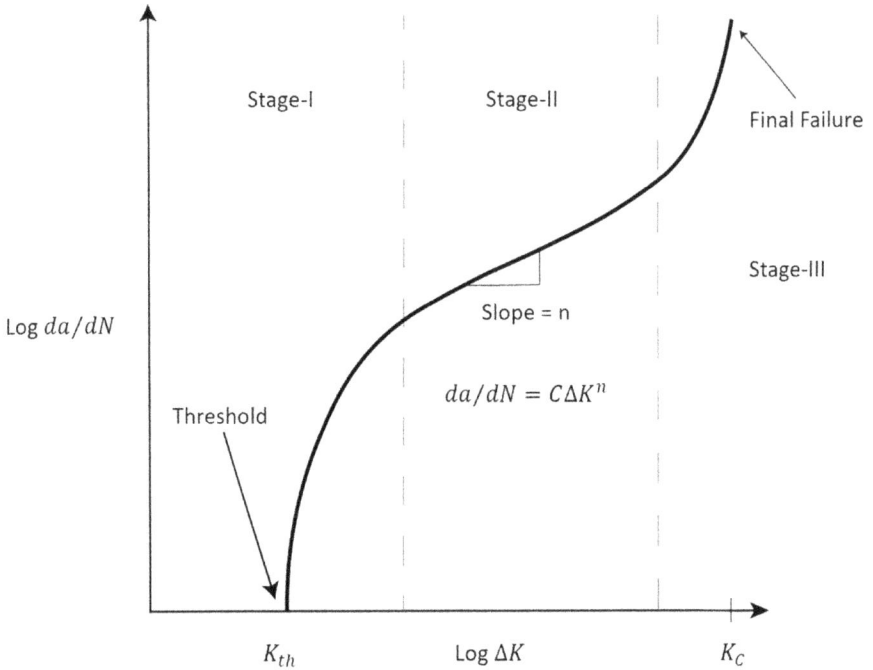

FIGURE 4.5 Fatigue crack growth rate curve typical in metals.

where C and n are material constants that are determined experimentally. Based on this equation, the growth rate of a fatigue crack is solely dependent on ΔK and the crack will continue to propagate at a steadily accelerating rate. In Stage III, the fatigue process continues to be influenced by the factors described in Stage I. However, with larger K values, the crack growth rate is higher, leading to greater instability and final fracture occurs when the stress intensity factor reaches the critical value K_C.

4.1.1.2 Fatigue Fracture Surface

There are usually two distinct regions that are visible: the fatigue crack propagation region (Stage II) and the final overload region (Stage III). The Stage II region is characterized by striation lines and benchmarks formed by the periodic loading and unloading cycles. These lines point to the direction of crack growth as shown in Figure 4.6.

In Stage III the crack propagates to failure with almost no plastic deformation, usually resulting in catastrophic failure. In this region the failure surface is smooth and void of striations. Typical fatigue fracture surfaces are shown in Figure 4.7

Concept Challenge 4.2

What is the difference between fatigue striations and benchmarks?

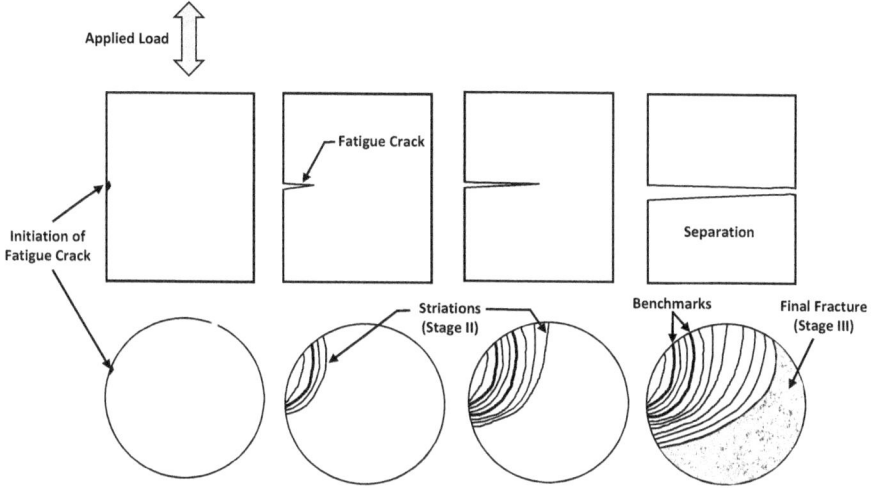

FIGURE 4.6 Schematic diagram of fatigue fracture surface showing various stages of crack propagation and striations.

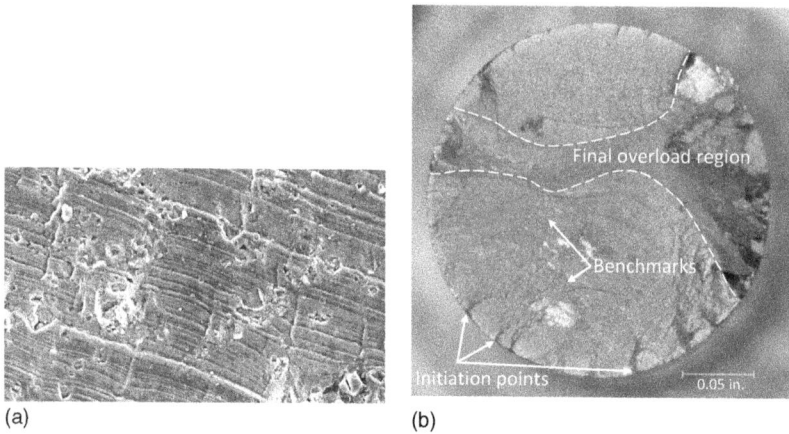

(a)

(b)

FIGURE 4.7 Fatigue fracture surfaces (a) Striation lines (b) Crack initiation location and final overload region on a fractured bolt surface [1].

Example 4.1

A large plate containing an edge crack is subjected to a constant loading frequency of 0.015 Hz during normal operations which causes the crack to grow from 3 to 7 mm. The maximum stress experienced during loading is 392 MPa and the stress ratio is 0.051. A similar plate material has a plane strain fracture toughness of 75 MPa√m and the crack growth can be described using the Paris law in the form

$$\frac{da}{dN} = 2 \times 10^{-12} (\Delta K)^4$$

where ΔK is in MPa$\sqrt{m}$. Determine (a) the maximum crack length the plate can support without failure, (b) the number of cycles before rupture, and (c) the time for rupture to occur.

It is assumed that the plastic zone due to ΔK is small and contained within the singularity dominated zone and that the plate is sufficiently large to be treated as an infinite plate.

(a)   From Eqn. (2.12) with $\alpha = 1.12$ and $\sigma_{max} = 392$ MPa, at the instant of fracture

$$K_{IC} = 1.12 \sigma_{max} \sqrt{\pi a_c}$$

$$75 = 1.12(392)\sqrt{\pi a_c}$$

$$a_c \approx 9.3 \text{ mm}$$

(b)   Since R is given, we can use Eqn. (4.1c) to determine σ_{min}

$$R = 0.051 = \frac{\sigma_{min}}{\sigma_{max}} = \frac{\sigma_{min}}{392}$$

$$\sigma_{min} \approx 20 \text{ MPa}$$

The stress range $\Delta\sigma$ is given by

$$\Delta\sigma = \sigma_{max} - \sigma_{min} = 372 \text{ MPa}$$

Using the Paris law equation

$$\frac{da}{dN} = 2\times10^{-12}\left(1.12\Delta\sigma\sqrt{\pi a}\right)^4$$

$$\int_{a_i}^{a_f} \frac{da}{a^2} = 2\times10^{-12}(1.12)^4(372)^4\pi^2\int_0^{N_f}dN$$

$$-(a)^{-1}\Big|_{a_i}^{a_f} = 59.48\times10^{-2}N$$

$$N = \frac{(0.003m)^{-1}-(0.0093m)^{-1}}{59.48\times10^{-2}\,m^{-1}\cdot cycles^{-1}} = 380 \text{ cycles}$$

(c)   The frequency f is 18 Hz and the time t to failure is given by

$$t = \frac{N}{f} = \frac{380 \text{ cycles}}{0.015 \text{ cycles/s}} = 25,333.3 \text{ sec}$$

$$t = 7.04 \text{ hours}$$

4.1.2 Industry Applications

4.1.2.1 Case Study 1: XFEM Simulation of Fatigue Crack Growth in a Welded Joint of a Pressure Vessel With a Reinforcement Ring Weldment [4]

A numerical analysis of the behavior of a pressure vessel with a reinforcement ring subjected to both static and dynamic loading is presented here. Pressure vessels normally contain various connections to pipelines such as a manhole to enable maintenance and repairs to be carried out when necessary. Such openings need to be adequately reinforced along their circumference. These connections are often made by welding different types of materials, which because of their heterogeneous nature along with the changes in the geometry lead to significant stress concentrations.

In this study, a crack was introduced to a welded joint connecting the reinforcement ring to the wall of the pressure vessel as shown in Figure 4.8, in order to determine its behavior under operation and test pressures. Pressure vessel walls were made from S275JO steel, whereas the reinforcement ring was made using steel P280 GH. Electrode EVB50 was used as filler material for welded joints. Two numerical models were developed to simulate crack growth under fatigue loading, to determine how different load conditions affect the integrity of the pressure vessel, in terms of crack behavior.

The numerical simulation of the first model was performed using the finite element method while the second used the extended finite element method (XFEM), which involves the use of so-called enrichment functions. The greatest advantage of XFEM compared to the classic finite element method is that there is no need to remesh the model during the simulation. Instead, the enrichment functions allow the finite elements to be used in a way that eliminates the discontinuities in the model, which are typically the consequence of cracks. XFEM application requires the use of special types of elements and nodes, which can be distinguished by the way in which the finite element nodes are generated. Two most frequently used types are the near-tip nodes and Heaviside nodes. An example of a crack represented by such finite elements is shown in Figure 4.9.

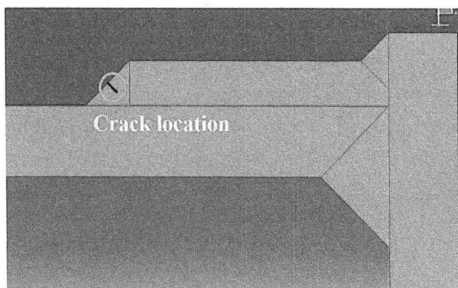

FIGURE 4.8 Location of a 5 mm crack in numerical model.

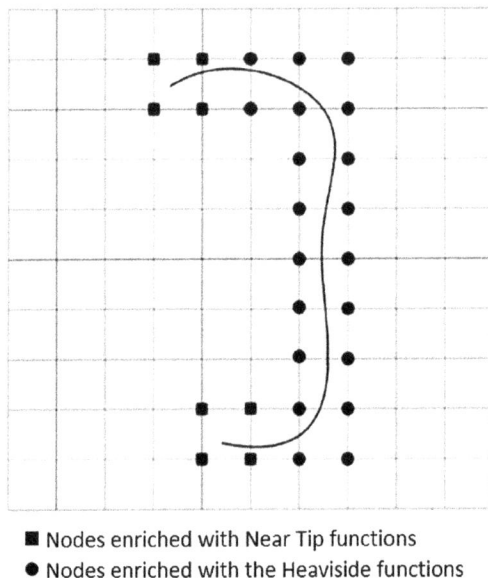

■ Nodes enriched with Near Tip functions
● Nodes enriched with the Heaviside functions

FIGURE 4.9 Extended finite element mesh using different types of enrichment functions.

ABAQUS software was used for both models. Morfeo software, an extension of ABAQUS used in fatigue simulations, was employed for the second analysis. For this reason, the second model had to be made three dimensional, unlike the first one which could be simulated in two dimensions. For the first model, the load was defined as a 50 bar (5 MPa) internal pressure, in accordance with the test pressure defined in the documentation. For the second model, the load was defined with the same magnitude, but was applied as a dynamic one, to simulate the fatigue behavior of the model. Morfeo performs its calculations based on the Paris equation $\left(\dfrac{da}{dN} = C\left(\Delta K\right)^{n} \right)$, with $C = 3.5$ and $n = 10^{-12}$ for the materials used. Simulation results for both models are shown in Figure 4.10.

The materials used for the first model were defined both in terms of elastic and plastic behavior, based on their mechanical properties. In the case of the second model, plastic behavior was excluded, since fatigue crack growth in this case was observed from the standpoint of linear-elastic fracture mechanics. In the case of the first model, the highest stresses were located at the connection of the bottom edge of the welded joint and the mantle, as well as around the crack tip, with magnitudes of 544.5 and 496.3 MPa, respectively (disregarding the stress values near the supports). The crack itself did not grow significantly, and the plastic strain in the model was minimal.

For the second model, the stresses reached noticeably higher levels, as expected, since the load was applied as a dynamic one. Stresses near the crack tip had exceeded 800 MPa, and the model ultimately failed after 56,290 load cycles. Due to the extreme load, the crack propagated through the welded joint and into the parent metal (P 280 GH),

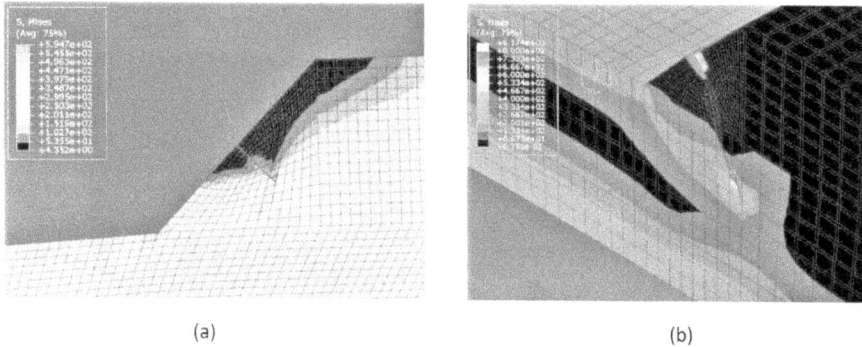

(a)   (b)

FIGURE 4.10 (a) Stress distribution for first model in welded joint area (b) Stress distribution and crack propagation in second (XFEM) model.

which was a result of the crack length and the weld metal geometry, as well as the fact that the WM (EVB50) is stronger than the parent material (yield strength of 440 MPa compared to 300 MPa, which is the yield strength of P 280 GH). The length of the crack in the final step was 17.678 mm.

The results obtained indicate that it is possible to quickly and efficiently compare models subjected to different load cases, although it should be mentioned that these models were also heavily approximated. Further work should focus on creating more realistic, optimized models that involve actual experimental determination of Paris coefficients C and m for the weld metal. Additionally, failure assessment diagrams should be made, after a more realistic fatigue load is adopted for future models, as drawing them in this extremely conservative model would not provide any significant insight into the behavior of the pressure vessel.

4.1.2.2 Case Study 2: Effect of Additional Holes on the Transient Thermal Fatigue Life of a Gas Turbine Casing [5]

Gas turbine casings are susceptible to cracking at the edge of an eccentric pin hole, which is the most likely position for crack initiation and propagation. In this study, the effect of adding additional holes to the casing is considered and the improvement of the transient thermal fatigue crack propagation life of gas turbine casings is discussed. Figure 4.11 is a 3D drawing of the casing showing the position of the eccentric pin hole.

The crack position and direction were determined using non-destructive tests. The gas turbine undergoes three processes; start up, base load, and shut down. The most important boundary conditions are temperature distributions on the inner and outer surfaces of the casing. The inner surface temperature was measured with ten type K thermocouples, within the range of 0–1,100 °C installed in ten different positions inside the casing. As a result of mechanical constraints, thermocouples were installed inside the retaining pin sites of the shrouds and nozzles, and the tip of the thermocouples penetrated inside the retaining pin site up to the casing's thickness. The basic pattern had one hole 5 mm in diameter (eccentric pin hole). A technical method for

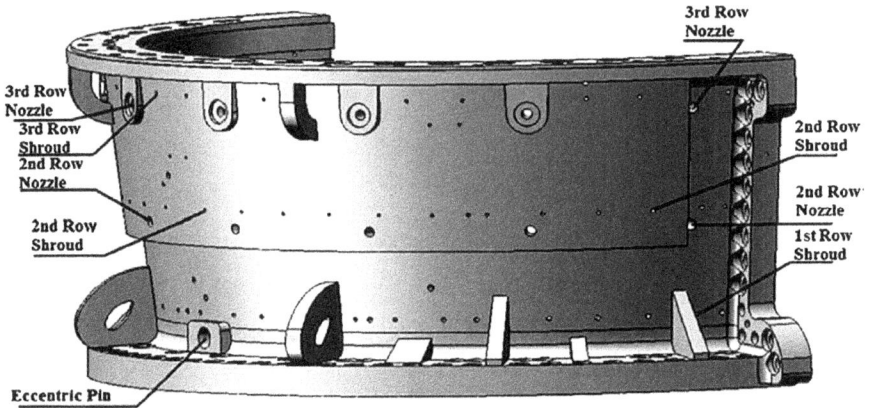

FIGURE 4.11 Turbine casing.

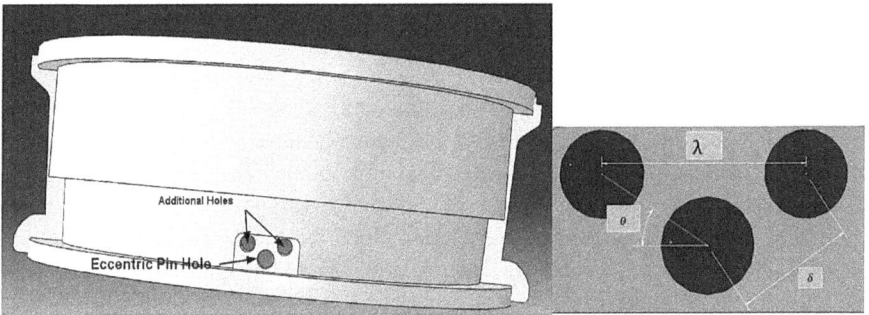

FIGURE 4.12 Arrangement of additional holes used in gas turbine casing.

improving the transient thermal fatigue crack propagation life was investigated by applying additional holes as shown in Figure 4.12.

Additional holes at different vertical distances apart were selected for various patterns with and without pins inserted into the holes. The vertical distance (λ) was taken as 150 mm or 200 mm, while the distance (δ) was set to either 100 mm, 125 mm, or 150 mm. Additional holes at different angles (θ) were also selected for 0°, 30°, 45°, and 60°.

A series of finite element patterns were developed and tested in ASTM-A395 elastic perfectly plastic ductile cast iron. The ABAQUS data file for uncracked mesh was read and processed by Zencrack using the energy release rate to drive the crack growth calculations. To calculate crack propagation, the well-known Paris law is used. The equation is:

$$\frac{da}{dN} = C\left(\Delta K\right)^{n}$$

where C and n are material properties. For Mode I loading fracture mechanics, the relationship between stress intensity factor K and energy release rate G is:

$$K = \left(\frac{EG}{1-(\alpha v)^2} \right)^{1/2}$$

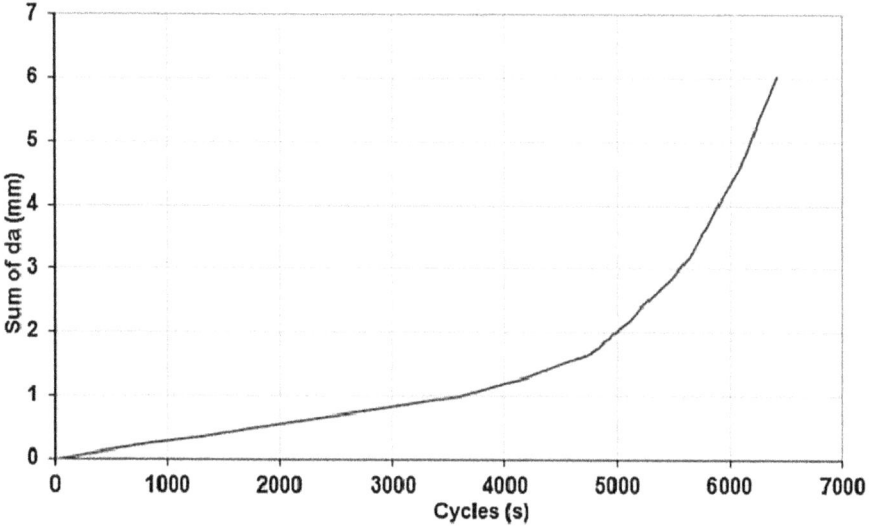

(a)

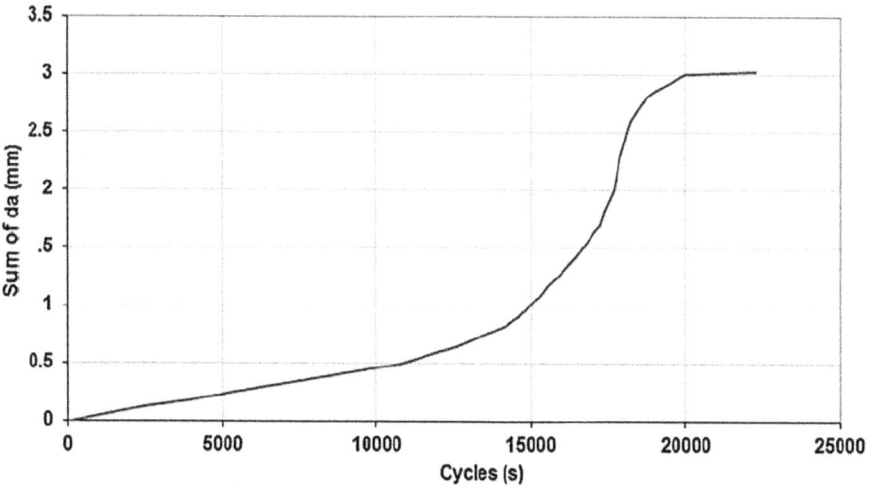

(b)

FIGURE 4.13   Transient thermal fatigue crack propagation life curve for casing (a) without additional holes and (b) with additional holes.

where α may be between 0 and 1 depending upon the state of stress ($\alpha = 0$ for plane stress to $\alpha = 1$ for plane strain). The best arrangement of additional holes was used to improve the transient thermal fatigue crack propagation life in a gas turbine casing. Figure 4.13 shows the fatigue life before and after the additional holes respectively.

The result shows that transient thermal fatigue crack propagation life could be extended by applying additional holes of larger diameter and be decreased by increasing the vertical distance, angle, and distance between the eccentric pin hole and the additional holes.

4.2   FAILURE ASSESSMENT DIAGRAMS

One of the most widely used methods for determining if a crack in a structural component will result in failure is the failure assessment diagram (FAD). The concept was introduced by Dowling and Townley [6] and focused on a two-criteria approach involving both elastic and plastic deformation to more accurately assess failure. The FAD approach can handle a variety of material behavior ranging from linear elastic brittle fracture to nonlinear ductile fracture. The Central Electricity Generating Board in the UK first published a procedure for failure assessment using the two-criteria approach in 1976 which became known as the R6 procedure [7]. Since then the procedure has undergone several revisions. A similar procedure was adopted by the British Standards Institution and subsequently issued as the BS 7910 standard of which the 2013 version [8] will be discussed here. In setting up a FAD, a relationship needs to be established between brittle fracture and plasticity. The first criterion requires that the stress intensity factor K_I does not exceed the fracture toughness of the material. This is accomplished by defining a dimensionless parameter K_r as:

$$K_r = \frac{K_I}{K_{mat}} \tag{4.7}$$

where K_I is defined using the appropriate expression for the stress intensity factor (SIF) from Table 2.1 and K_{mat} is the linear elastic fracture toughness (K_{IC}) for brittle fracture or an equivalent toughness (computed using the J-integral or crack tip opening displacement (CTOD)) when there is significant plastic deformation which will be discussed shortly.

The second criterion, the load ratio L_r, is a measure of the structure's proximity to plastic collapse and is given by:

$$L_r = \frac{P}{P_c} \tag{4.8}$$

where P is the applied load and P_c is the plastic limit load in the cracked component. L_r can also be expressed as a stress ratio

$$L_r = \frac{\sigma_{ref}}{\sigma_c} \tag{4.9}$$

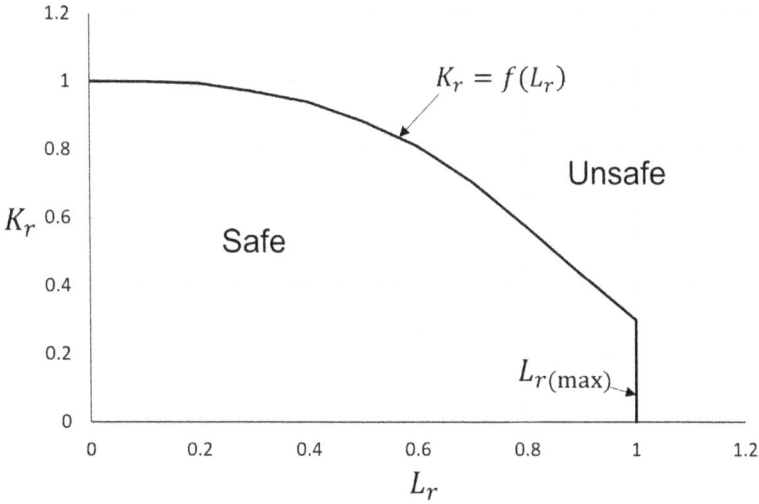

FIGURE 4.14 Typical failure analysis diagram.

where σ_{ref} is the applied stress and σ_c is the stress required to cause plastic collapse of the structure. For tensile loading, this collapse occurs when the stress on the net cross-section causes the material to deform at a constant strain rate in its plastic region and is also known as the flow stress of the material. This means that σ_c is dependent not only on the material properties but also on the flaw size. The last step in creating the FAD is the introduction of a limiting load, $L_{r(max)}$, on the horizontal axis which acts as a cut-off value. A typical FAD curve plotted using the general relationship $K_r = f(L_R)$ is shown in Figure 4.14.

The BS 7910 standard outlines three options for creating a FAD. Option 1 is the simplest and does not require stress-strain data while options 2 and 3 are more complicated and require additional material data and stress analysis but produce increasingly accurate results. In the case of option 1, the function defining K_r for $L_r \leq L_{r(max)}$ is given by:

$$K_r = \left[1 - 0.14(L_r)^2\right]\left\{0.3 + 0.7 \exp\left[-0.65(L_r)^6\right]\right\}$$ (4.10)

The FAD represents an assessment envelope that is bounded by the axes, the curve defined by K_r, and the cut-off value $L_{r(max)}$ which is defined as

$$L_{r(max)} = \frac{\sigma_{ys} + \sigma_u}{2\sigma_{ys}}$$ (4.11)

where σ_u is the ultimate tensile strength. If an assessment point (L_r, K_r) is plotted on the FAD and lies within this envelope, the flaw is considered acceptable. The transition between brittle fracture, elastic-plastic deformation, and plastic collapse can be seen in Figure 4.15.

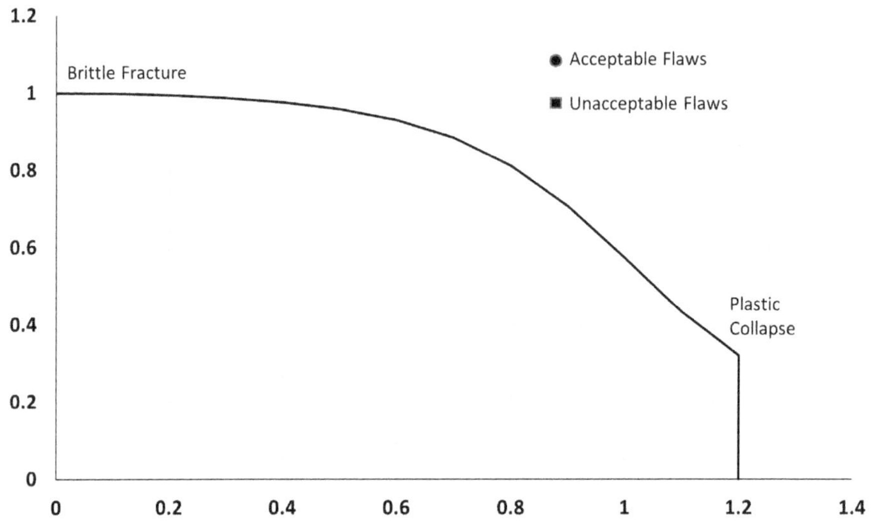

FIGURE 4.15   Failure analysis diagram showing evaluation points.

Example 4.2

A 1020 mild steel panel is subjected to an applied load of 1.5 MN. The panel is 500 mm wide with a thickness of 10 mm and contains a 90 mm central crack. The material has a σ_{ys} = 330 MPa, σ_u = 441 Mpa, and fracture toughness K_{mat} = 200 MPa$\sqrt{m}$. Using the FAD shown in Figure 4.16 determine whether this panel will fail or not.

The cut-off stress is found using Equation (4.11):

$$L_{r(max)} = \frac{\sigma_{ys} + \sigma_u}{2\sigma_{ys}} = \frac{(330 + 441)MPa}{2(330)MPa} = 1.168$$

To determine the L_r coordinate, we will need to estimate the flow stress σ_f, which is responsible for plastic collapse:

$$\sigma_f = \frac{\sigma_{ys} + \sigma_u}{2} = \frac{(330 + 441)MPa}{2} = 384.5\,MPa$$

The stress on the net cross-section is given by

$$\sigma_{ref} = \frac{(1.5 \times 10^6)N}{(0.01m)(0.5m - 0.09m)} = 364.9\,MPa$$

and

$$L_r = \frac{341.5\,MPa}{384.5\,MPa} = 0.949$$

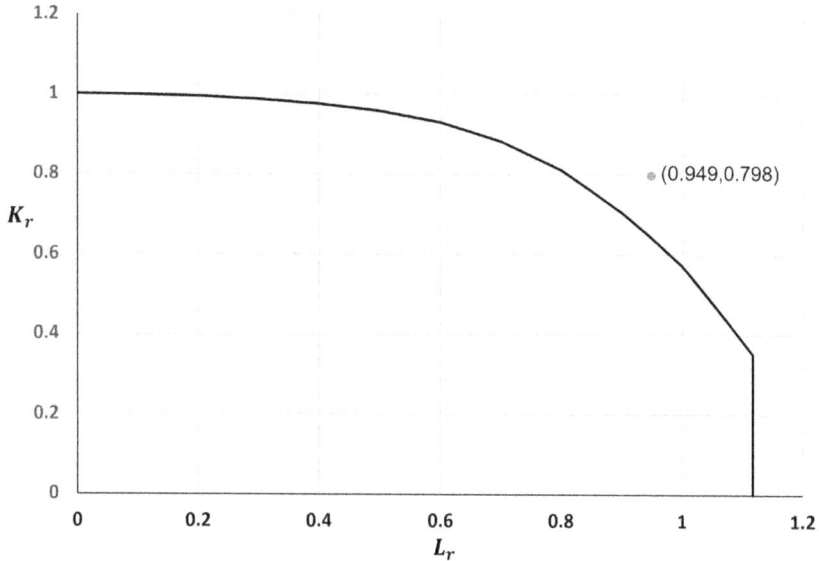

FIGURE 4.16   Graph showing evaluation point (0.949, 0.798) plotted on a FAD.

Using Eqn. (2.16) and ignoring the higher order terms (since $\dfrac{a}{W}$ is small):

$$K_I = \frac{P}{B\sqrt{W}}\sqrt{\frac{\pi a}{4W}}\sec\left(\frac{\pi a}{2W}\right)$$

$$= \frac{\left(1.5\times10^6\right)}{0.010m(\sqrt{0.25}}\sqrt{\frac{\pi\left(0.09m\right)}{4\left(0.25m\right)}}\sec\left(\frac{\pi\left(0.09m\right)}{2\left(0.25m\right)}\right)$$

$$= 159.5\,MPa\sqrt{m}$$

The K_r coordinate is given by

$$K_r = \frac{K_I}{K_{mat}} = \frac{159.5\,MPa\sqrt{m}}{200\,MPa\sqrt{m}} = 0.798$$

Since the point lies outside of the curve, the flaw is unacceptable and the panel will fail before reaching the applied 1.5 MN load.

Concept Challenge 4.3

(a) If a brittle fracture analysis alone was performed in Example 4.2, would the panel fail?

(b) What if the analysis was solely based on collapse?

4.3 APPLICATIONS IN FAILURE ANALYSIS

So far we have considered several techniques that are useful in allowing us to predict when failure is likely to occur in an attempt to avoid catastrophic events. However, even if the design is sound, failure may still occur for a variety of reasons including misuse or improper maintenance. In such cases, the analysis of the failure becomes an extremely important aspect of engineering in terms of determining the cause of failure. The process offers the engineer insight into possible design improvements and a better understanding of operating procedures and conditions for which the part was designed. Although a complete treatment of failure analysis is not possible in this section we will consider some of the more common types of engineering failures and their related mechanisms.

4.3.1 FUNDAMENTALS OF FAILURE ANALYSIS

With the exception of corrosion, failure of engineering materials generally occurs as a result of fracture due to overload in tension, torsion, bending, or fatigue conditions. Fracture in a brittle material typically occurs along a plane that is normal to the principal stress direction, whereas ductile materials, which tend to flow more readily, exhibit fracture along maximum shear stress planes.

Macroscopic examination of the fracture surface under low magnification (< 20x) can provide useful details in deducing the mode of loading responsible for fracture surface topology. This is accomplished by comparing the characteristic features observed in the failed specimen with macrofractographic data for known fracture conditions. The material type, microstructure, and loading conditions of the in-service component affects the appearance of the fracture surface.

4.3.1.1 Tensile Overload

Figure 4.17 illustrates a resulting fracture surface for cylindrical specimens of both brittle and ductile materials that are subjected to overload conditions in simple tension. In the case of a brittle specimen, the maximum normal stress develops along the

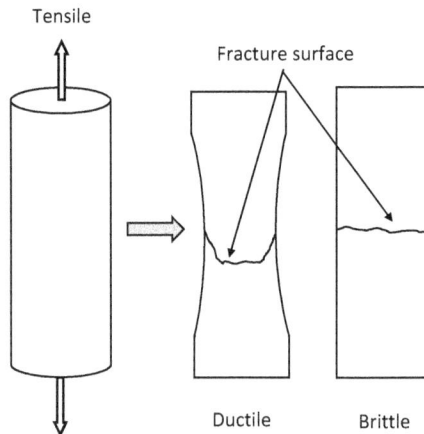

FIGURE 4.17 Depiction of fracture surfaces for ductile and brittle cylindrical specimens loaded in simple tension.

vertical-axis of the cylinder and fracture occurs on a plane normal to this axis. For a ductile specimen, the mechanism of failure becomes a bit more complex and involves two mechanisms. The central portion develops a plane strain condition due to the Poisson effect on the surrounding material which serves to limit plastic deformation, resulting in a brittle fracture surface normal to the direction of loading.

Since the outer surface of the cylinder is traction free, as we move past the central region, the material begins to flow more easily and large scale plastic deformation is allowed to occur. In this region the maximum shear stress prevails and the fracture surface changes orientation to 45^0. Fractographs of both brittle and ductile fracture surfaces are as shown in Figure 4.18. The radial lines correspond to the plane stress region.

In the case of a tensile loaded ductile specimen that has a rectangular cross-section, such as a plate or sheet, radial marks develop around the long axis of the specimen as shown in Figure 4.19.

These are known as chevron or herringbone marks. It can be observed from the fractograph in Figure 4.20 that these lines point back toward the origin of fracture. As the specimen thickness decreases, the chevron line formation tends to become more symmetrical about the midline of the plate.

4.3.1.2 Torsion Overload

When the cylindrical specimen composed of a brittle material is subjected to a simple torsional load, the maximum normal stresses occur at 45^0 to the vertical-axis. The resulting crack travels along a spiral path producing a fracture surface that is inclined at 45^0. In a ductile specimen, the maximum shear stress occurs instead at 90^0 (Figures 4.21

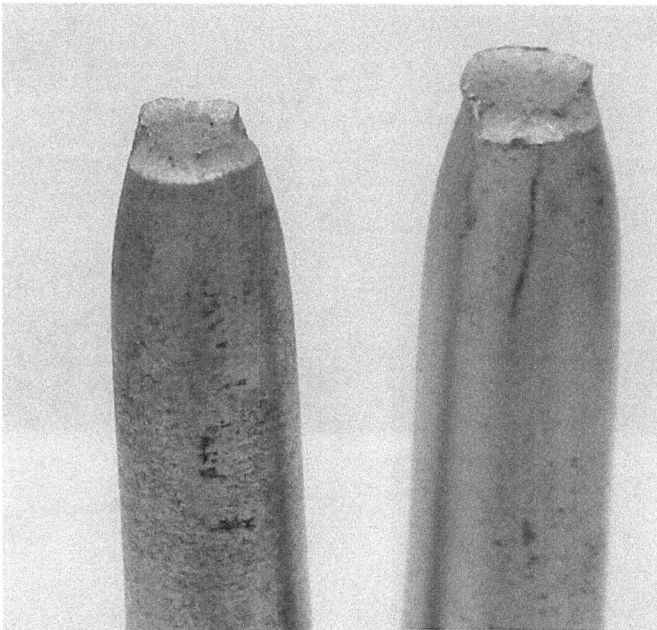

(a)

FIGURE 4.18 Fractographs of (a) ductile

(Continued)

(b)

FIGURE 4.18 (CONTINUED)   Fractographs of (b) brittle specimens resulting from tensile overload failure [9].

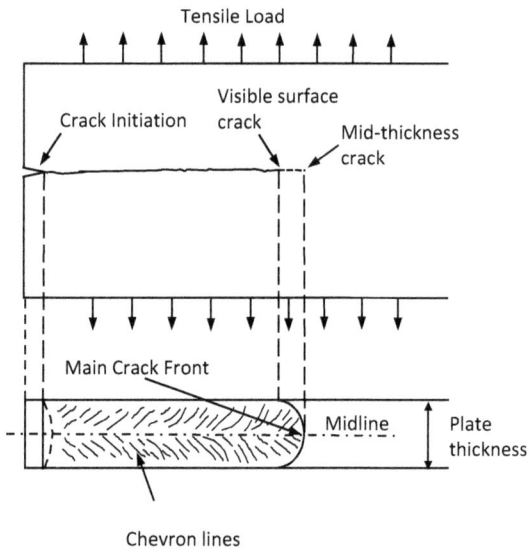

FIGURE 4.19   Formation of chevron lines on fracture surface of a thin plate.

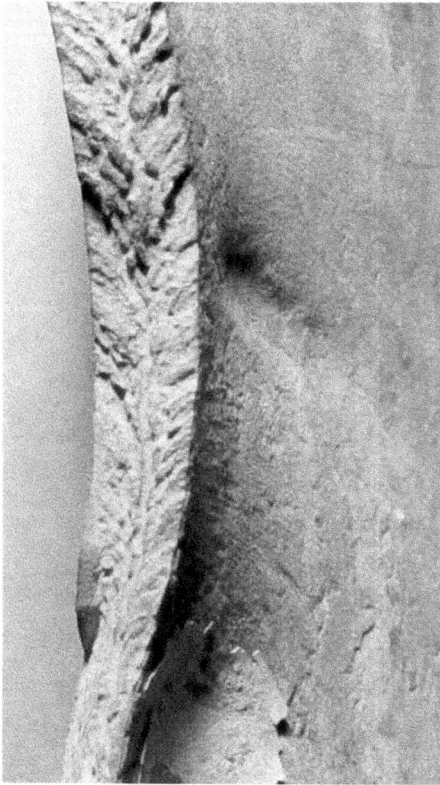

FIGURE 4.20 Fracture surface showing chevron marks in shell of steel tank [10].

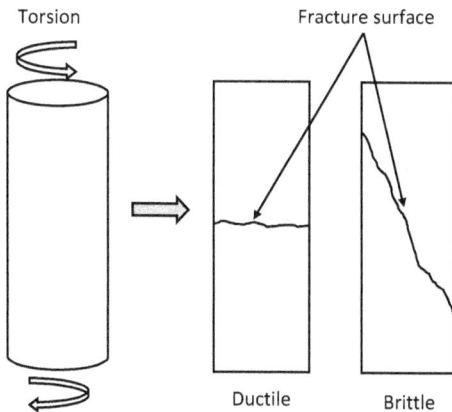

FIGURE 4.21 Depiction of fracture surfaces of ductile and brittle cylindrical specimens loaded in simple torsion.

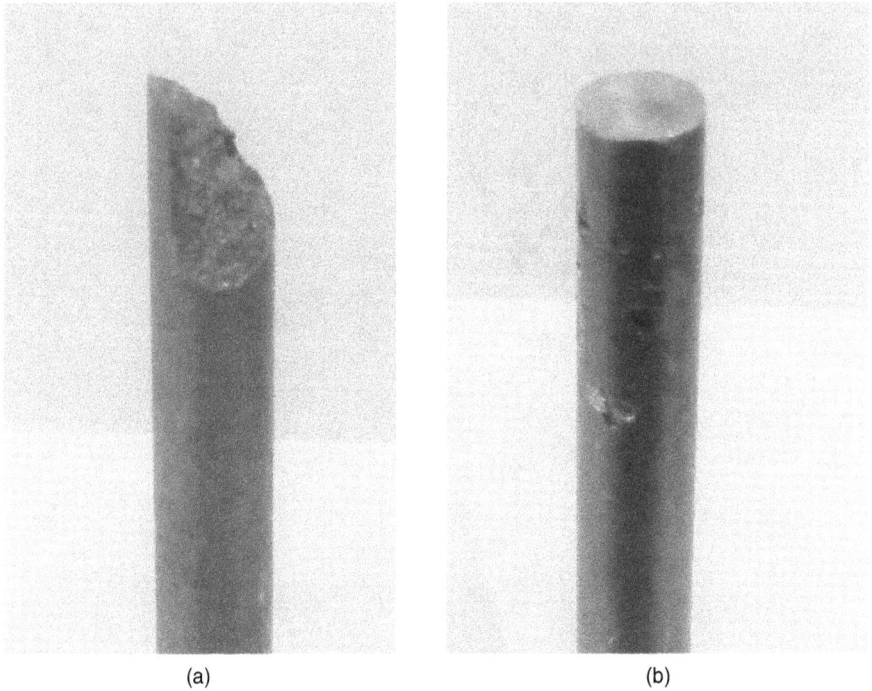

(a)   (b)

FIGURE 4.22   Fracture surface resulting from torsion overload for (a) ductile and (b) brittle materials.

and 4.22). The material is able to flow around the cylindrical axis due to its ductile nature, resulting in a swirl-like fracture surface.

4.3.1.3   Bending Overload

For overload in simple bending, the maximum normal stress occurs parallel to the vertical-axis. The stress value ranges from tension on the outer surface, gradually decreasing to zero as we approach the center of the cylinder, before switching to compression as we move to the opposite surface. Cracks form on the tensile side of the specimen and propagate within the cylinder until separation occurs. With a brittle specimen, the fracture surface is normal to the cylinder axis, while in a ductile specimen it is along the 45^0 plane as shown in Figure 4.23.

The fractographs in Figure 4.24 show these characteristic features for brittle and ductile materials resulting from a bending overload failure.

4.3.2   Industry Applications

4.3.2.1   Case Study 3: Failure of a 40-inch Diameter Crude Oil Pipeline [12]

In August 2009, there was a 2.5-meter-long rupture in the longitudinal seam weld of a crude oil pipeline in France (Figure 4.25). The failure caused a spillage

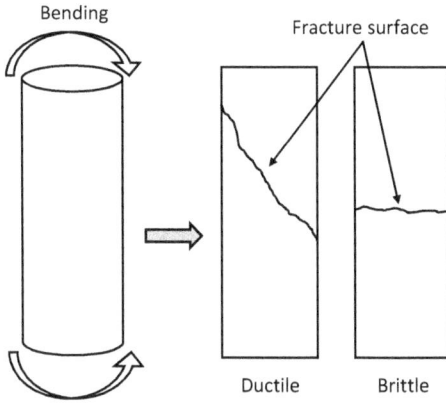

FIGURE 4.23 Depiction of fracture surfaces for ductile and brittle cylindrical specimens loaded in simple bending.

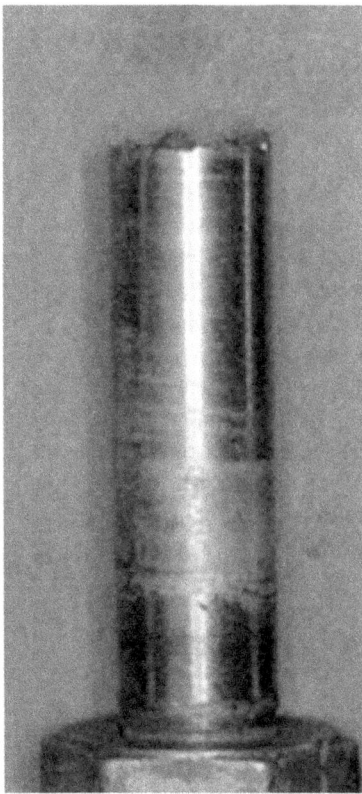

(a)

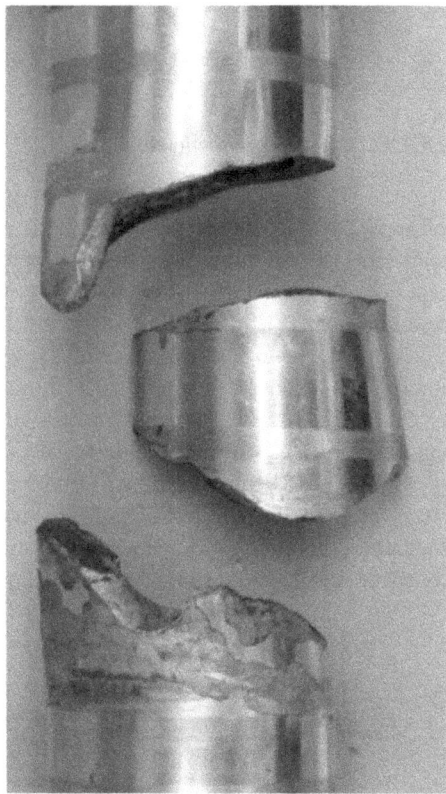

(b)

FIGURE 4.24 Fracture surface resulting from bending overload for (a) ductile [11] and (b) brittle materials [3].

FIGURE 4.25   Failed section of pipe.

of approximately 2,000 cubic meters in a protected area. This rupture caused the authorities to withdraw the permit to operate a 260-kilometer-long section of this pipeline. Penspen Ltd was contracted by the pipeline operator Société du Pipeline Sud Européen (SPSE) to carry out an independent investigation into the cause of the failure, review and confirm the actions needed for safe short-term operation to allow internal inspection, and determine a safe future life for the pipeline.

Initial analytical work indicated that the failure may have been related to the presence of a crack-like seam weld defect in 1972, when the pipeline went into operation, or it might have been caused by a combination of cyclic loading and roof topping as shown in Figure 4.26. Roof topping is caused by failure to crimp the plate edges sufficiently in the pipe process and may result in a decrease in fatigue life of a pipeline by concentrating local stresses in the seam weld.

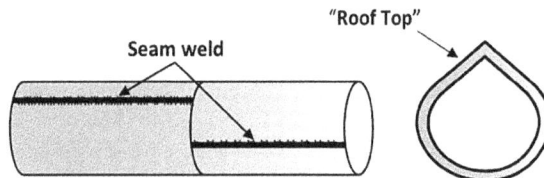

FIGURE 4.26   Illustration of the seam weld and roof-topping effect.

A program of small-scale material tests and ring testing were conducted to quantify the burst and fatigue performance of the pipeline; this testing was carried out by GL Noble Denton (GLND). The ring tests, of line pipe cut from the pipeline, included 11 static ("burst") tests and 13 fatigue tests. For the static tests one specimen was defect free while the others contained electric discharge machined slits (EDM) to represent crack-like defects.

The key objective of these static tests was to be able to predict "critical defect sizes" – defects that would cause failure at the pipeline's operating pressure. A secondary objective was to understand the failure behavior, that is whether the failure was predominantly ductile or brittle. The fatigue tests were performed in order to understand the development and growth of fatigue cracks at the longitudinal seam weld toe. Eleven fatigue tests had initial EDM slits and two did not.

The results of this testing program were analyzed in order to investigate the cause of previous failures, and estimate the pipeline's remaining fatigue life. A two-step approach to analyze the test results and explain the failure involved (i) S-N analysis using empirical data obtained from fatigue tests and (ii) an analytical model based on fracture mechanics to model seam-weld fatigue-crack growth from an initial crack size to the critical size. It should be noted that fatigue is not a major cause of failure in oil and gas pipelines, but when it does occur, it is usually associated with longitudinal seam-weld defects, or damage such as dents. The study revealed that the cause of failures of the SPSE pipeline in France was a combination of fatigue due to pressure cycling of the pipeline, roof topping, and a pre-existing defect which may have possibly been present when the pipeline went into service in 1972.

4.3.2.2   Case Study 4: Failure Study of The Railway Rail Serviced for Heavy Cargo Trains [13]

In this case study, a failed railway rail which was used for heavy cargo trains is investigated in order to find out its root cause. In addition to the fatigue load, rails are also subjected to other high mechanical loads and harsh environmental conditions. The main loading components are rolling contact pressure, shear and bending forces from the vehicle weight, thermal stresses due to restrained elongation of continuously welded rails, and residual stresses from manufacturing and welding in the field. The latter makes it more difficult to control in terms of weld quality, due its dependence on the operator. In this study, a failed railway rail shown in Figure 4.27, which was used for heavy cargo trains, was investigated in order to find out its root cause. The analysis involved performing macroscopic inspections, chemical analysis, scanning electron microscope (SEM) observations, and metallographic examinations.

From the macroscopic failure morphology shown in Figure 4.28, it was found that the fracture surface was basically clean and fresh, which demonstrated that the corrosion of the fracture surface was not heavy. Since the railway rail was subjected to cyclic loading and had served about six years, it is rational to consider that this railway rail might have failed due to fatigue. The arc boundary of the fan-shaped area looks like a beach mark when observed macroscopically. The crack origin might be at the corner of the darkly fan-shaped area. However, taking into account that the

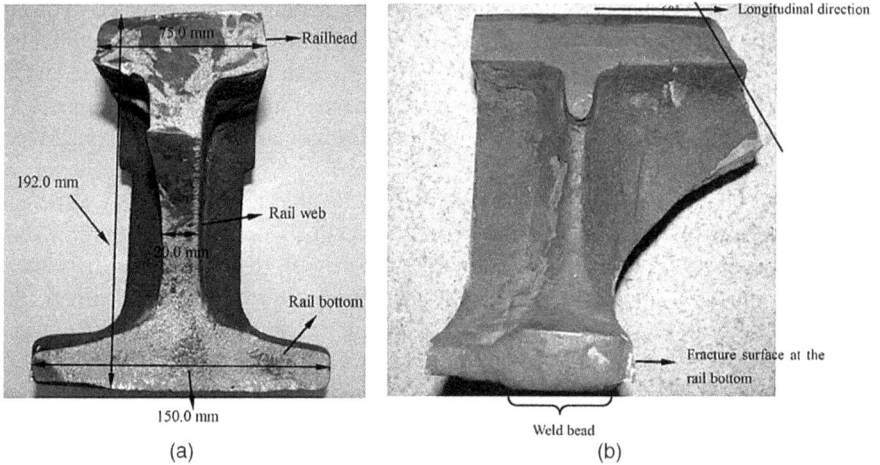

FIGURE 4.27 Failed railway rail. (a) Front view. (b) Lateral view.

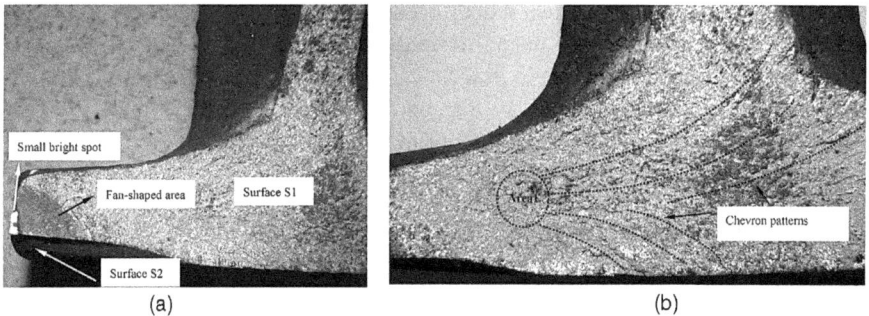

FIGURE 4.28 Macroscopic morphologies of the failed railway. (a) Rail bottom. (b) Chevron patterns and crack origin at the fracture surface of the rail bottom.

small bright spot was next to this darkly fan-shaped area, it can be deduced that this would not be the crack origin. The beach marks which were the classical features of metal fatigue were not observed from the macroscopic observations. However, the chevron patterns can be clearly observed at the fracture surface of the rail bottom. Therefore, the crack origin should be at the tip of the chevron patterns and the crack growth direction is along the diverging direction of the river patterns. It can be deduced from the flat fractography and chevron patterns that the macro-fracture feature is a brittle fracture.

Although the railway rail was mainly subjected to cyclic loading, the macroscopic beach marks and microscopic fatigue striations were not observed at the fracture surface. In addition, the typical chevron and cleavage fracture was observed at the tip of the chevron patterns. After complete analysis, we can conclude that the failure of the railway rail was mainly caused by overload, even though it was subjected to cyclic loading.

4.4   NON-DESTRUCTIVE TESTING

In many cases, it is not always possible to assess the integrity of an in-service engineering structure or component in the laboratory using standard testing procedures
that may result in permanent damage to the sample. Non-destructive testing (NDT)
collectively refers to a group of techniques that allow for the inspection, testing, or
evaluation of components or structures for defects, discontinuities, or differences in
characteristics without affecting their functionality or service life [14]. In addition to
in-service inspection, NDT can also be used during manufacturing and fabrication processes to ensure product integrity and reliability as well as the construction phase for
maintaining consistent quality in processes such as welding and joining of materials.

Several industries utilize NDT methods including aviation, automotive, oil and
gas, power generation, naval, and aerospace. Although NDT does not eliminate the
risk of failure, it can certainly mitigate it. In this section we consider some of the
more common NDT methods, particularly those that relate to the detection of cracks
or defects such as voids that can lead to fracture failure.

4.4.1   ULTRASONIC TESTING

Ultrasonic testing (UT) continues to be one of the most effective methods of modern
NDT. This method involves inducing high-frequency sound waves into solid objects,
typically metals or composites, using a transducer that converts electrical energy
into ultrasonic waves [15]. The travel of these waves is affected when they encounter irregularities such as density variations, cracks, voids, honeycombs, or foreign
objects. A receiver is used to collect these waves that are either reflected back to the
source or that pass through the material being inspected as shown in Figure 4.29. The
UT equipment can then map the interior of the object by analyzing and interpreting
the returned sound wave. The most common sound frequencies used in UT range
from 1.0 to 10 MHz.

While traditional UT employed a single transducer, modern equipment involves
phased array ultrasonic testing (PAUT) that relies on several transducers operating
synchronously. This greatly increases the inspection speed and widens the coverage

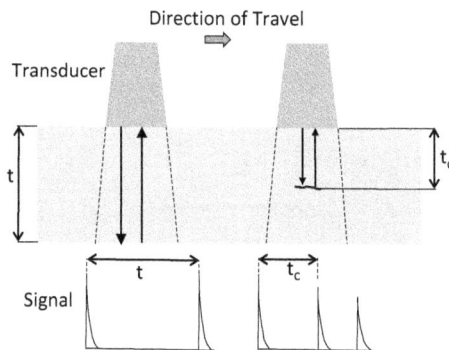

FIGURE 4.29   Crack detection using a single ultrasonic transducer.

area. PAUT also produces very accurate readings and requires a minimal set-up time with the test equipment being light and portable. Although there are several advantages to UT, it is not without limitations. Certain materials with large grain sizes such as iron, austenitic steels, and welds can interfere with the wave transmission causing attenuation which can mask defects. In addition, odd geometries such as curved or rough surfaces can lead to alignment issues and inaccurate results. UT also requires a skilled operator, and training can be costly and extensive when compared to other NDT methods.

4.4.2 EDDY CURRENT TESTING

Eddy current testing (ECT) is based on the principle of electromagnetic induction. When an alternating current flows through a conductor, it produces a magnetic field which in turn generates a small secondary current known as an eddy current [16]. The flow pattern of this current is affected when it encounters discontinuities within the test piece as shown in Figure 4.30. ECT uses the change in eddy current density to produce images of the conductive material and is capable of detecting corrosion, voids, crack delamination, and loss of thickness. ECT is popular with several industries such as power generation, aerospace, rail, oil and gas, and manufacturing due to its portability, speed, and accuracy.

ECT works best on smooth surfaces for surface and near-surface defects due to its limited penetration capability, usually less than 0.25 inches. Although the eddy currents can penetrate thin nonconductive coatings, such as the zinc coating on galvanized steel sheets, it is otherwise limited to conductive materials. Inspection of ferromagnetic materials can also pose a challenge due to magnetic permeability.

4.4.3 MAGNETIC FLUX LEAKAGE

The magnetic flux leakage (MFL) method is used to detect anomalies in normal flux patterns created by discontinuities in ferrous material saturated by a magnetic field

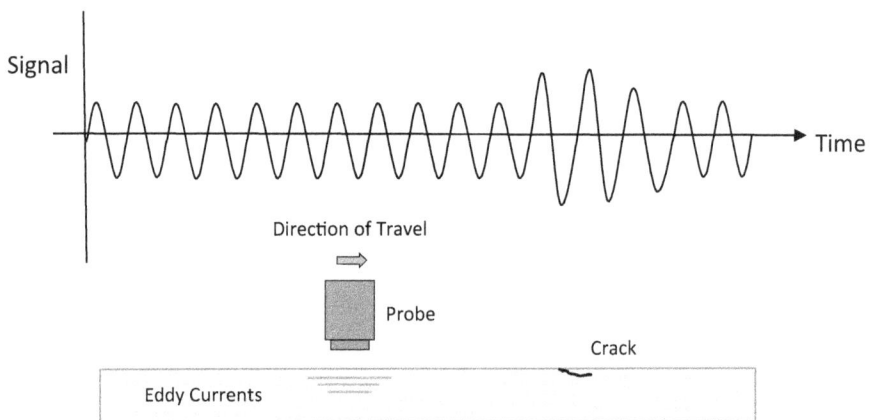

FIGURE 4.30 Crack detection using ECT method.

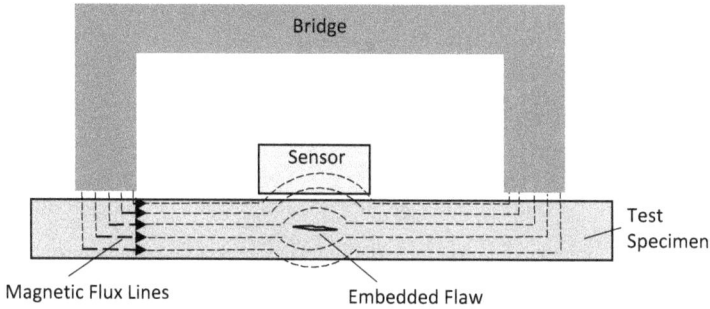

FIGURE 4.31   Flaw detection using the MFL method.

created by a powerful magnet [17]. A sensor is used to detect fluctuations in the magnetic field caused by differences in material properties. This technique can be used for piping and tubing inspection, tank floor inspection, and other similar applications. It can be done without removing the insulation, resulting in a fast, economic way to inspect long runs of pipe or tubing. MFL is very effective at detecting surface and subsurface flaws. However, it is limited to ferromagnetic materials only and produces very poor results when detecting axial cracks or deep flaws.

Figure 4.31 shows an application of MFL for a tank floor inspection using a series of magnetic field generators known as "bridges" and sensors located side by side across the front of the machine. The bridges generate a magnetic field that saturates the tank floor allowing the detection of cracks, thickness losses, pitting, or corrosion.

4.4.4   RADIOGRAPHIC TESTING

Industrial radiographic testing (RT) involves exposing the test sample to penetrating radiation that passes through the object being inspected to a receiving medium on the opposite side of the object. The emerging radiation can be processed using film radiography, computed radiography, computed tomography, or digital radiography [18]. X-rays are normally used for thinner or less dense materials such as aluminum, while gamma rays are used for thicker or denser materials such as structural steels.

Regardless of the medium used, the radiation will show flaws in the material based on the strength of the radiation reaching the detection medium. In the case of film radiography, darker areas are produced where more radiation has passed through the specimen and lighter areas where less radiation has penetrated. If there is a void or defect present, more radiation passes through, causing a darker image on the film or detector, as shown in Figure 4.32.

The RT method produces permanent records of the defects, and test equipment is easily portable. It is also capable of measuring or detecting internal defects, porosity, inclusions, cracks, lack of fusion, corrosion, geometry variation, density changes, and part misalignments. The technique does have some drawbacks in terms of the radiation hazard as well as being costly. It also requires both sides of the part to be accessible and does not provide the depth of the defect.

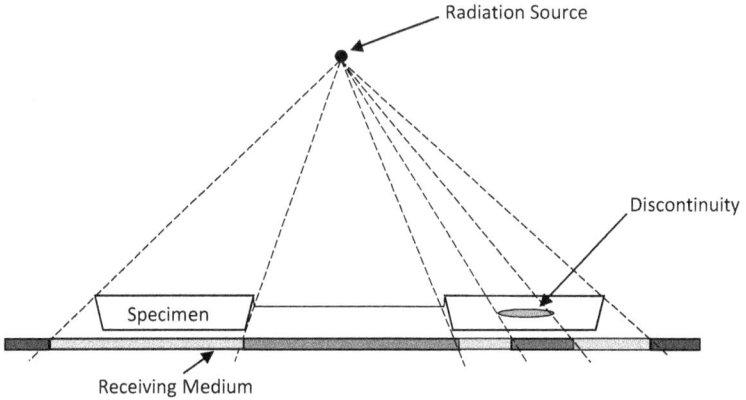

FIGURE 4.32   Flaw detection using the RT method.

4.4.5   LIQUID PENETRANT TESTING

Liquid penetrant testing (PT) involves the application of a very low viscosity (highly fluid) liquid, known as the penetrant, to the surface of the part to be tested. Prior to the liquid application, the surface should be cleaned and free of oil, grease, water, or other contaminants. Once the fluid is applied it seeps into any defect such as cracks, fissures, and voids that are open on the material surface [19]. Sufficient time is needed for this step to allow as much penetrant as possible to be drawn into the defect. The excess penetrant is then removed from the surface while removing as little from the defect as possible. Penetrants may be visible in ambient light, or fluorescent, requiring the use of an ultraviolet light. The process using a visible liquid as penetrant is shown in Figure 4.33. A light coating of developer is then applied to the surface to draw up any penetrant trapped in the defect back to the surface again where it will be visible.

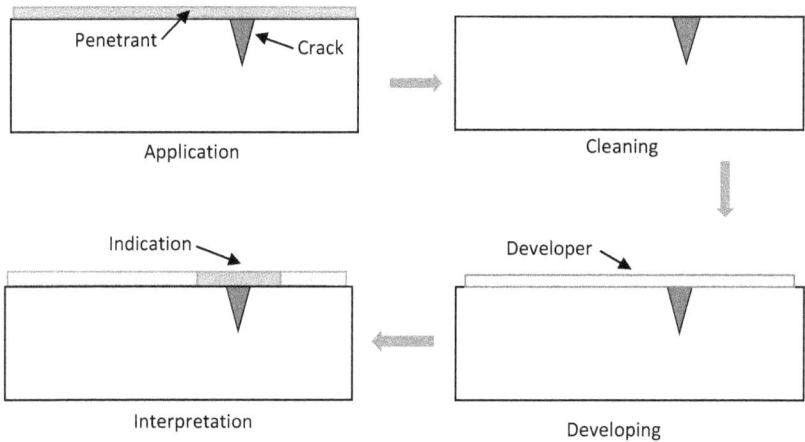

FIGURE 4.33   Crack detection using PT method.

The developer is allowed to stand for the prescribed dwell time before inspecting the part visually, with appropriate lighting in order to detect any flaws that may be present. The part surface must be thoroughly cleaned to remove the remaining developer.

The use of PT has several advantages including high sensitivity that allows small discontinuities to be detected, rapid inspection of large areas, portability, low cost, and is suitable for parts containing complex geometries. It can also be used on metallic, nonmetallic, magnetic, nonmagnetic, conductive, and nonconductive materials. Although versatile, PT is only capable of detecting surface breaking cracks and does not work well on porous materials. Surface finish and roughness can affect sensitivity. Pre- and post-cleaning is also necessary and proper disposal of chemicals is required.

PROBLEMS

4.1. A fatigue test was conducted in which the mean stress was 50 MPa and the alternating stress was 215 MPa. Compute (a) the maximum and minimum stress values, (b) the stress ratio, and (c) the stress amplitude.

4.2. A fatigue test on an aluminum alloy gives the following results:

Stress Amplitude (MPa)	Cycles to Failure
375	1×10^4
300	5×10^4
250	1×10^5
200	1×10^6
160	1×10^7
150	5×10^7
146	1×10^8

(a) Create an S-N plot for the data (stress amplitude versus logarithm cycles to failure) and

(b) using the plotted data determine the fatigue strength at 5×10^5 cycles.

4.3. In the Paris law given by Eqn. (4.9), what is the physical meaning of the slope m for the Stage II portion of the crack growth curve?

4.4. A gas turbine disk experiences stress amplitudes of 500 MPa during normal operations. Inspection reveals the presence of a crack of length $a = 55$ μm. If the standard service life for the disc is 16,000 cycles, determine the largest allowable size for the defect if the crack growth relationship for the material is given by

$$\frac{da}{dN} = 3.86 \times 10^{-10} (\Delta K)^3$$

4.5. A gas turbine housing contains an internal crack of half length $a = 200$ μm. Normal operation produces a stress amplitude of 800 MPa and the critical half crack length of the material was found to be 2 mm. Determine the life of the housing if the crack growth rate, da/dN, (m/cycle) is related to the alternating stress intensity, ΔK (MPa$\sqrt{m}$) by:

$$\frac{da}{dN} = 5 \times 10^{-12} (\Delta K)^3$$

4.6. A large part subjected to a cyclic load of $\Delta\sigma = 275$ MPa with a stress ratio of zero behaves according to the following Paris law:

$$\frac{da}{dN} = 4.25 \times 10^{-7} (\Delta K)^{2.5}$$

where da/dN is in m/cycles and ΔK is in MPa$\sqrt{m}$. Determine the plane strain fracture toughness required for the part to endure 8,340 cycles, if the initial crack length is 2 mm.

4.7. The table below shows data obtained (BS 7910 standard procedures) for a wide sheet containing a central crack.

K_r	0.00	0.63	0.71	0.74	0.77	0.79
L_r	0.52	0.54	0.57	0.64	0.77	0.96

The FAD envelope can be generated using the following data:

$K_r = f(L_r)$	0	0.2	0.3	0.4	0.6	0.8	0.9	1.0	1.0
L_r	1	0.99	0.98	0.97	0.92	0.8	0.72	0.6	0.0

By plotting the (L_r, K_r) data and the FAD envelope, determine the critical crack size resulting in catastrophic failure. The material has $\sigma_{ys} = 414\,MPa$; $\sigma_f = 320\,MPa$; $K_{mat} = 130\,MPa\sqrt{m}$ and the fracture toughness is given by:

$$K_{IC}^2 = \frac{8}{\pi}\sigma_{ys}^2 \, a \ln\left[\sec\left(\frac{\pi\sigma_f}{2\sigma_{ys}} \right) \right]$$

REFERENCES

[1] R. O. Richie, "Mechanisms of fatigue-crack propagation in ductile and brittle solids," *International Journal of Fracture*, vol. 100, pp. 55–83, 1999.
[2] P. C. Paris and F. Erdogan, "A Critical Analysis of Crack Propagation," *Journal of Basis Engineering, Transactions of ASME*, vol. 85, pp. 528–534, 1963.
[3] R. Hicks, "Material Laboratory Factual Report No. 09-045," National Transportation Safety Board, Washington, 2009.

[4] E. Dzindo, S. A. Sedmak, A. Grbovic, N. Milovanovic and B. Dodrevic, "XFEM simulation of fatigue crack growth in a welded joint of a pressure vessel with a reinforcement ring," *Archive of Applied Mechanics*, vol. 89, pp. 919–926, 2019.

[5] H. Bazvandi, "Effect of additional holes on transient thermal fatigue life of gas turbine casing," *Case Studies in Engineering Failure Analysis*, vol. 9, pp. 78–86, 2017.

[6] A. R. Dowling and C. H. Townley, "The Effect of Defects on Structural Failure: A Two-Criteria Approach," *International Journal of Pressure Vessel and Piping*, vol. 3, pp. 77–137, 1975.

[7] R. P. Harrison, K. Loosemore and I. Milne, "Assessment of the Integrity of Structures Containing Defects," *CEGB Report R/H/R6, Central Electricity Generating Board, UK*, 1976.

[8] BS 7910:2013 - "Guide to methods for assessing the acceptability of flaws in metallic structures", British Standards Institution, 2013.

[9] J. D. Glassman, A. Gomez, M. E. M. Garlock and J. Ricles, "Mechanical properties of weathering steels at elevated temperatures," *Journal of Constructional Steel Research*, vol. 168, pp. 1–10, 2020.

[10] J. Henderson, "Materials Laboratory Draft Factual Report No. 05-071," National Transportation Safety Board, Washington, 2005.

[11] D. Jones, "Materials Laboratory Factual Report No. 13-062," National Transportation Safety Board, Washington, 2013.

[12] M. Dafea, P. Hopkins, R. Palmer-Jones, P. d. Bourayne and L. Blin, "An investigation into the failure of a 40-in diameter crude oil pipeline," *The Journal of Pipeline Engineering*, no. 1st Quarter, 2014.

[13] Y. D. Li, C. B. Liu, N. Xu, X. F. Wu, W. M. Guo and J. B. Shi, "A failure study of the railway rail serviced for heavy cargo trains," *Case Studies in Engineering Failure Analysis*, vol. 1, pp. 243–248, 2013.

[14] M. Rucka, "Non-Destructive Testing of Structures," *Materials*, vol. Special Issue, 2020.

[15] ASTM, "E164-13 Standard Practice for Contact Ultrasonic Testing of Weldments," ASTM International, 2013.

[16] ASTM, "E2884-17, Standard Guide for Eddy Current Testing of Electrically Conducting Materials Using Conformable Sensor Arrays," ASTM International, 2017.

[17] ASTM, "E570-20, Standard Practice for Flux Leakage Examination of Ferromagnetic Steel Tubular Products," ASTM International, 2020.

[18] ASTM, "E94 / E94M-17, Standard Guide for Radiographic Examination Using Industrial Radiographic Film," ASTM International, 2017.

[19] ASTM, "E1417 / E1417M-16, Standard Practice for Liquid Penetrant Testing," ASTM International, 2016.

5 Further Fracture Mechanics Applications

OBJECTIVES

After studying this chapter, the student should be able to:

1. Explain the safe-life, fail-safe, and damage tolerance approaches for failure prevention.
2. Explain the concepts of inspectability, and slow crack growth vs. fail-safe structures.
3. Explain the concepts of crack growth retardation and residual strength.
4. Apply fracture mechanics (FM) to determine inspection intervals.
5. Explain the leak before burst (LBB) concept.
6. Solve elementary LBB problems.

5.1 DESIGN APPROACHES TO PREVENT FAILURE

Documented analysis of a scientific approach to fatigue failure can be found as early as 1842 with the investigation of the Versailles rail accident in Meudon, France. The axle on the locomotive broke while in service resulting in 55 fatalities. The investigator, William John Macquorn Rankine, discussed the impact of cyclical loading on areas where stresses were magnified due to changes in geometry (stress concentration factors) [1]. Twentieth century approaches to preventing structural failure due to cyclical loading can be categorized as safe-life and fail-safe. These are described below.

5.1.1 SAFE-LIFE

In the *safe-life* approach, a useful service life is estimated (with a safety factor) based on component testing. The tests use load conditions similar to or which exactly match typical service load spectra. This approach does not consider the damage mechanisms that lead to failure, such as fatigue crack growth; its emphasis is therefore on damage initiation. The safe-life approach is therefore based on the assumption that the structure is initially defect free. The component is replaced at the end of its predicted life, even if failure has not occurred. The standard safe-life approach assumes the availability of a relevant S-N curve and applies a cumulative damage model such as the Palmgren–Miner rule (described in the next paragraph) to calculate the fatigue life. A process schematic of the safe-stress-life approach is shown in Figure 5.1. Note that a safe-strain-life approach can also be adopted.

The Palmgren–Miner rule defines the damage D_i accumulated from a constant amplitude cyclical stress as the ratio of the number of cycles, n_i, at the applied stress level to the number of cycles to failure N_{fi} at the same stress level. The total damage

DOI: 10.1201/9781003052050-5

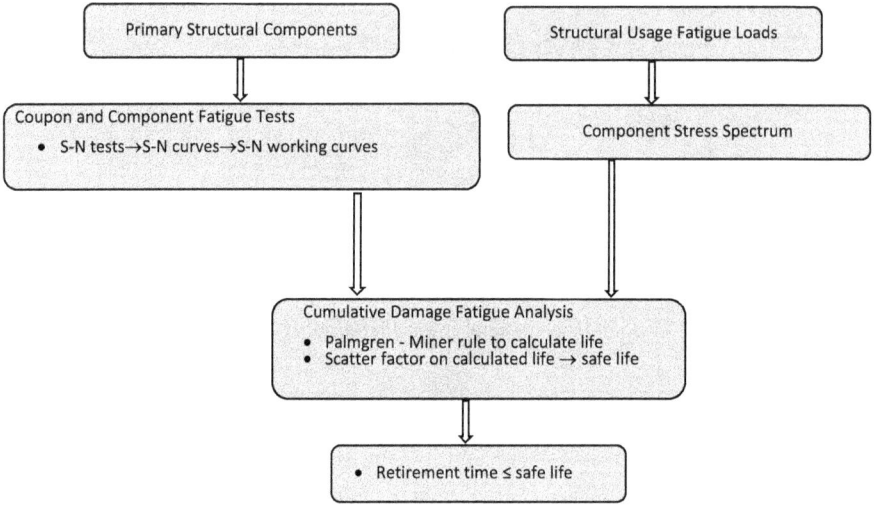

FIGURE 5.1 Safe-stress-life approach using the S-N curve.

is then a summation of the ratios at the various stress cycles as shown in Eqn. (5.1). Figure 5.2 shows an example of a load spectrum with two load cycles; failure will occur if $D = 1$.

$$D = \sum_{i=1}^{k} \frac{n_i(\sigma_i)}{N_{fi}(\sigma_i)} \qquad (5.1)$$

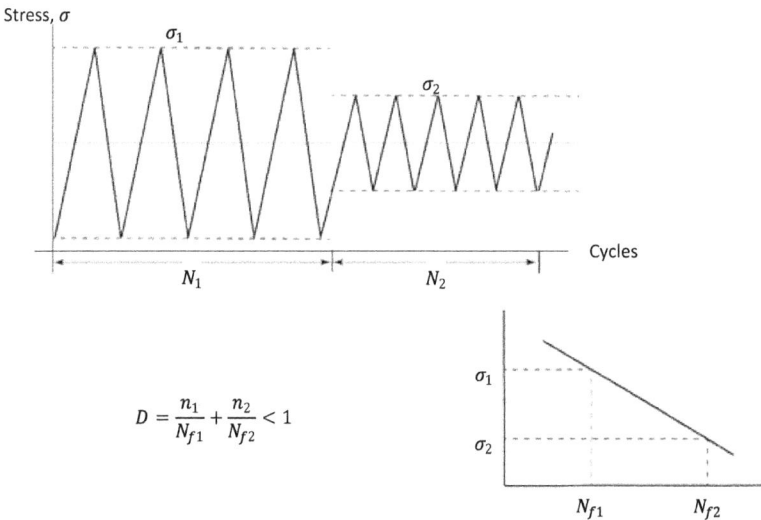

$$D = \frac{n_1}{N_{f1}} + \frac{n_2}{N_{f2}} < 1$$

FIGURE 5.2 Palmgren–Miner rule with two applied stress cycles.

5.1.2 FAIL-SAFE APPROACH

In the *fail-safe* approach, structures are designed with redundant systems, therefore if one member fails, there are enough alternative load paths available such that the structural integrity can be maintained, at least until the failed member can be identified and repaired. Periodic inspections which use damage detection methods capable of identifying flaws are mandatory. If damage is detected, parts are repaired or replaced promptly. These structures are called *fail-safe* structures. Load path redundancy is not a new concept. In the 15th century, Leonardo da Vinci wrote, "In constructing wings one should make one cord to bear the strain and a looser one in the same position so that if one breaks under strain the other is in position to serve the same function" (Leonardo da Vinci's notebook on the design of "Flying machines").

5.1.3 FAIL SAFE VS. SAFE LIFE

Safe-life designed components tend to be heavier and therefore more expensive than fail-safe designed components, therefore it is often cost prohibitive to have all components designed using a safe-life approach. Additionally, since the safe-life approach does not consider fracture mechanisms, it is sometimes subject to catastrophic failure if a crack develops during service, or an initial crack existed in a high stress and/or geometrically vulnerable location, which eventually grew to the point of instability during service. An initial crack can be caused by a host of phenomena, including but not limited to thermal stress cycles during fabrication, residual stresses due to interference fit, mechanical impact during production, or inherent material imperfections such as vacancies. In some industries, critical safe-life components are still inspected at certain frequency intervals. For example, landing gear are critical and though designed using the safe-life approach are still subjected to periodic inspections. In civil engineering, the fail-safe category is generally preferred for economic reasons, although there are exceptions [2].

5.2 DAMAGE TOLERANCE ANALYSIS

Damage tolerance analysis (DTA), developed in the 1970s, focuses on crack growth and safety by inspection. It is primarily used in the aerospace industry. DTA assumes cracks are always present in the structure and that they will grow due to fatigue and corrosion; however, the process can be understood and controlled. Emphasis is of course placed on structurally critical components, which are assumed to contain a crack in the high stress and/or high cycle regions. The process then uses FM to determine appropriate inspection frequencies and methods for cracks and component fatigue lives. Material properties (e.g. K_{Ic}, $\dfrac{da}{dN}$ vs. ΔK) and expected or known in-service loads are required inputs for the analysis. In conventional aircraft design, major wing joints, hinges on all-moving tailplanes or variable geometry wings, and wing-fuselage joints are designed using a safe-life approach; wing and fuselage skins stiffened by stringers and frames are designed using a DTA approach.

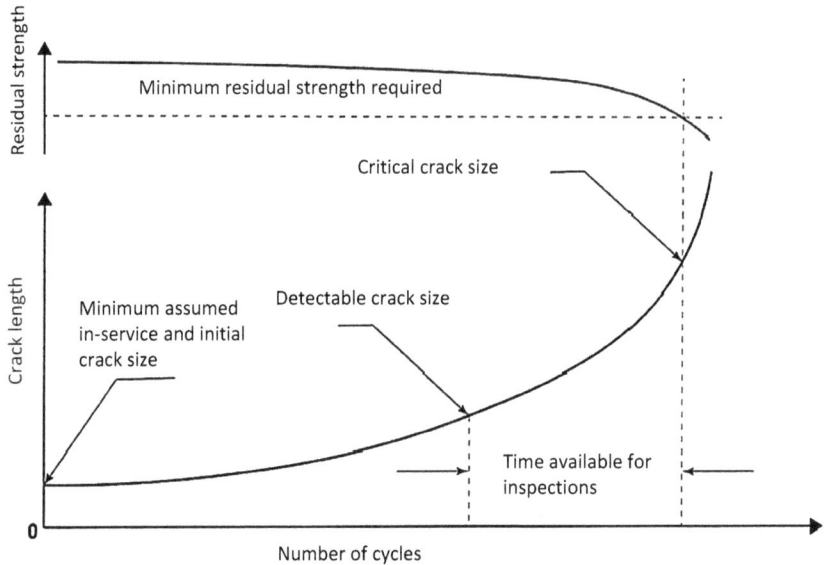

FIGURE 5.3   The damage tolerance approach.

The Federal Aviation Administration (FAA) develops, manages and enforces the Federal Aviation Regulations or FARs which govern all aviation activities in the United States. FAR 25.571 amendment 25-72 describes the difference between fail safe and damage tolerance as follows

> "Fail-safe generally means a design such that the airplane can survive the failure of an element of a system or, in some instances one or more entire systems, without catastrophic consequences.
>
> Damage-tolerance requires an inspection program tailored to the crack progression characteristics of the particular part when subject to the loading spectrum expected in service. Damage-tolerance places a much higher emphasis on these inspections to detect cracks before they progress to unsafe limits, whereas fail-safe allows cracks to grow to obvious and easily detected dimensions."

DTA defines a safe growth period (a period of unrepaired service usage) based on the design life requirement for the component and/or scheduled in-service inspection intervals. The *residual strength* is defined as the static load that the structure can sustain without failure, in the presence of a crack. Residual static strength generally decreases with increased damage size. Figure 5.3 shows crack size increasing as residual strength decreases with an increasing number of cycles. The time available for inspection is limited by some minimum residual strength requirement.

5.2.1 SAFETY ASSURANCE SLOW CRACK GROWTH VS. FAIL SAFE

DTA qualifies a structure within either a *slow crack growth* or a *fail-safe* category. A slow crack growth structure is centered around a strict adherence to disallowing flaws or defects from attaining the critical size required for unstable rapid crack propagation. Safety assurance arises from the fact that the crack is growing slowly and the specified periods of usage and degrees of inspectability are appropriate [3].

For the fail-safe category, the design includes redundant load paths and crack growth arrest mechanisms to negate overall failure when a propagating crack achieves its critical length under service loads. It is therefore expected that unstable rapid crack propagation will stop prior to complete failure as loads are increasingly transferred to adjacent structures. Safety assurance arises from the slow crack growth of the remaining structure and the detection of the damage at subsequent inspections [3].

5.2.2   RESIDUAL STRENGTH CURVE

The residual strength curve is plotted using the stress intensity vs. crack length relationship unique to the geometry and loading of concern. This was discussed in Chapter 2 and is repeated here for convenience.

$$K_I = Y\sigma\sqrt{\pi a} \tag{5.2}$$

When $K_I = K_{Ic}$, $\sigma = \sigma_{crit}$, therefore

$$K_{Ic} = Y\sigma_{crit}\sqrt{\pi a} \tag{5.3}$$

σ_{crit} vs. a can now be plotted if K_{Ic} is known. This is demonstrated in Example 5.1.

Example 5.1

Plot the residual strength curve for a center cracked infinite panel made from Al 7075-T6 ($K_{Ic} = 27MPa\sqrt{m}$). If it is subjected to a tensile stress of 50 MPa when the crack length ($2a$) is 4 cm, will it fail? What would the crack length have to be for failure to occur when the plate is subjected to a 50 MPa tensile load? What would the stress value need to be for failure to occur when the crack length is 5 cm?

SOLUTION

For a center cracked infinite plate under tension $Y \approx 1$. Therefore

$$\sigma_{crit} = \frac{27}{\sqrt{\pi a}}$$

The plot is shown in Figure 5.4. The point (0.02 m, 50 MPa) is below the residual strength curve, the plate will therefore not fail. If we extend the horizontal and vertical lines from this point to the curve, the intersections yield the critical crack size and critical stress respectively. Therefore, the critical crack length at 50 MPa is estimated at 18.4 cm ($a = 0.092$ m); the critical stress at a 4 cm crack length is estimated at 107 MPa. At a crack length of 5 cm ($a = 0.025$), the critical stress can be read directly from the curve and is approximately 96.4 MPa.

Residual Strength Curve for Al 7075-T6

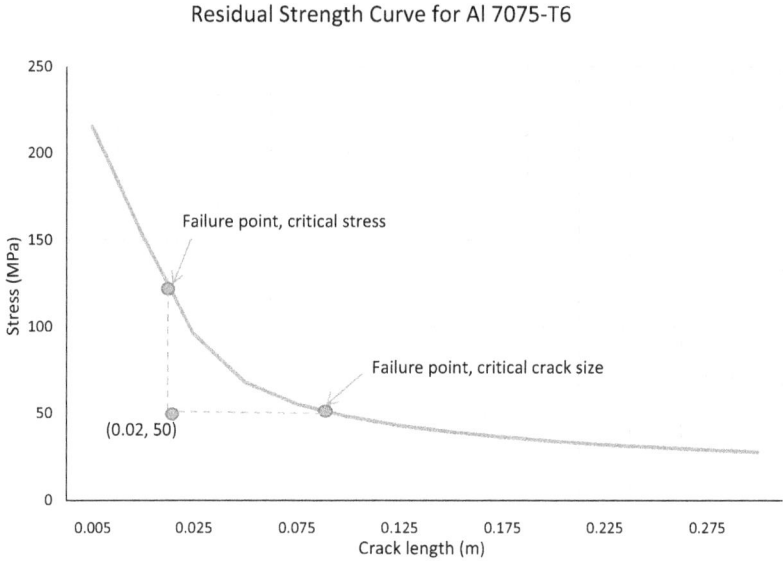

FIGURE 5.4   Residual strength curve for center-cracked panel Al 7075-T6.

Note that *Y* varies with *a/W*, where *W* is the specimen width; therefore, that variation would also have to be accounted for when plotting the residual strength curve for a more realistic scenario.

5.2.3   INSPECTABILITY

Both commercial and military aircraft are subject to various levels of inspectability at intervals that correspond to the rigor of the inspection. The levels for the military are described below. These definitions are taken directly from the Department of Defense Joint Service Specification Guide, JSSG-2006 paragraph 6.1.15. Table 5.1

TABLE 5.1
Summary of In-Service Inspections from JSSG-2006 Appendix Table X

Degree of Inspectability	Typical Inspection Interval
In-flight evident inspectable	One flight[a]
Ground evident inspectable	One day (two flights)[a]
Walkaround inspectable	Ten flights[a]
Special visual inspectable	One year
Depot or base level inspection	A quarter of design service lifetime
In-service non-inspectable structure	One design service lifetime

[a] Most damaging mission

shows the related inspection intervals for each level of inspectability. Safety is assured through slow crack growth for specified periods of usage depending upon the degree of inspectability.

- **Depot or base level inspectable:** if the nature and extent of damage will be detected utilizing one or more selected nondestructive inspection procedures. The inspection procedures may include nondestructive inspection (NDI) techniques such as dye penetrant, X-ray, ultrasonic, etc. Accessibility considerations may include removal of those components designed for removal. **In-flight evident inspectable:** if the nature and extent of damage occurring in flight will result directly in characteristics which make the flight crew immediately and unmistakably aware that significant damage has occurred and that the mission should not be continued.
- **In-service non-inspectable structure:** if either damage size or accessibility preclude detection during one or more of the above inspections.
- **Ground evident inspectable:** if the nature and extent of damage will be readily and unmistakably obvious to ground personnel without specifically inspecting the structure for damage.
- **Special visual inspectable:** if the nature and extent of damage is unlikely to be overlooked by personnel conducting a detailed visual inspection of the aircraft for the purpose of finding damaged structure. The procedures may include removal of access panels and doors and may permit simple visual aids such as mirrors and magnifying glasses. Removal of paint, sealant, etc. and use of NDI techniques such as penetrant, X-ray, etc., are not part of a special visual inspection.
- **Walkaround inspectable:** structure is walkaround inspectable if the nature and extent of damage is unlikely to be overlooked by personnel conducting a visual inspection of the structure. This inspection normally shall be a visual look at the exterior of the structure from ground level without removal of access panels or doors without special inspection aids.

5.2.4 CRACK GROWTH RETARDATION

In FM, the term "retardation" refers to the reduction in crack growth rate after an overload has occurred in a sequence of low amplitude cycles. During tensile loading, the plastic zone enlarges, causing the elastic region just beyond the plastic zone to also deform. When the tensile load is released, the surrounding (elastic) material at the plastic–elastic interface will want to return to its original shape thereby causing a part of the plastic zone to experience compressive stresses. The larger the load, the larger the zone of compressive stresses. These residual compressive stresses can resist additional crack opening, particularly if the subsequent loads are much lower than the initial causal tensile load.

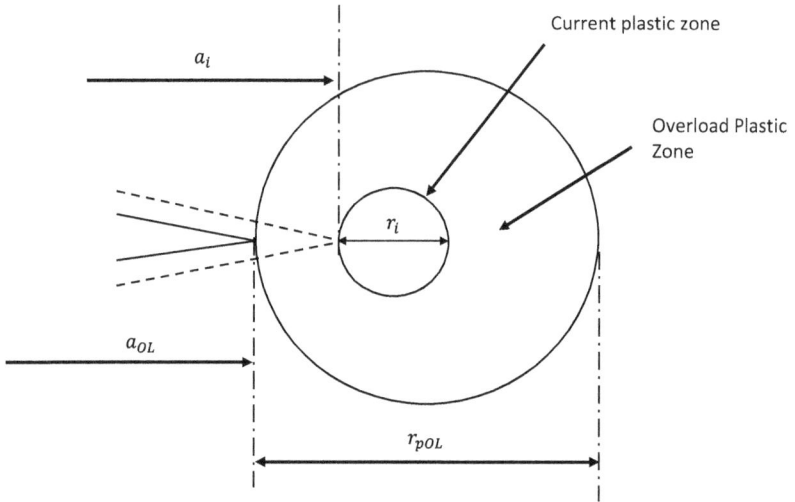

FIGURE 5.5   Crack growth reduction parameters for the Wheeler model.

5.2.5   The Wheeler Retardation Model

There are several models for crack retardation. The Wheeler model [4] is one of the simpler accepted models and will be discussed here. Wheeler defines a retarded or reduced crack growth rate $\left(\dfrac{da}{dN}\right)_r$ such that

$$\left(\frac{da}{dN}\right)_r = C_p f\left(\Delta K\right) \tag{5.4}$$

where $f(\Delta K)$ is the usual crack-growth function, and C_p is the crack growth rate reduction factor
C_p is given as

$$C_p = \left(\frac{R_{pi}}{a_{OL} + r_{poL} - a_i}\right)^m \tag{5.5}$$

The parameters in Eqn. (5.5) are described below and shown in Figure 5.5
R_{pi}, current plastic zone size in the ith cycle under consideration
r_{poL}, plastic zone size generated by overload
a_{OL}, crack size at which the overload occurred
a_i, current crack size
m, empirical constant

5.2.6   Initial Steps in the Damage Tolerance Process

The damage tolerance process has several steps [3] and can quickly increase in complexity. The first few steps are presented here as an introduction.

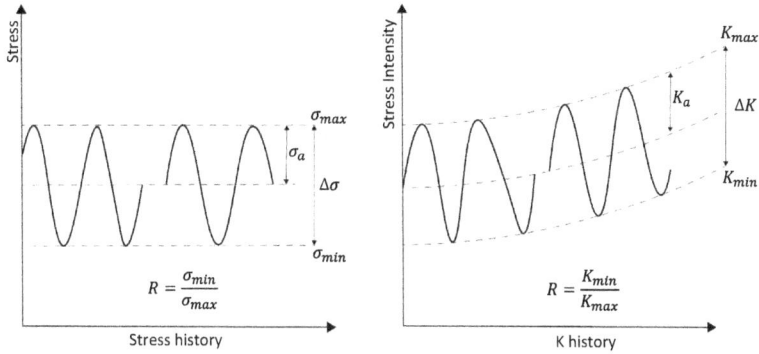

FIGURE 5.6 Sample stress history.

TABLE 5.2
Stress Intensity History Equations

The maximum stress intensity factor, K_{max}	$K_{max} = Y\sigma_{max}\sqrt{\pi a}$
The minimum stress intensity factor, K_{min}	$K_{min} = Y\sigma_{min}\sqrt{\pi a}$
The mean stress intensity factor, K_m	$K_m = Y\sigma_m\sqrt{\pi a}$
The amplitude of the stress intensity factor, K_a	$K_a = Y\sigma_a\sqrt{\pi a}$
The range of the stress intensity factor, ΔK	$\Delta K = Y\Delta\sigma\sqrt{\pi a}$
The cycle ratio, R_K	$R_K = K_{max}/K_{min}$

1. Determine the stress-intensity factor (K_I) as a function of crack size for the relevant part. The Y associated with the unique geometry and loading may be found in fracture mechanics tables, texts, or software databases.
2. Obtain the stress history for the location under consideration. This may be from data collected during service or generated from a simulation model using service loads. The stress history can then be converted into a stress intensity factor history at a given crack length by applying Eqn. (5.2), resulting in the equations in Table 5.2. A sample stress history conversion is shown in Figure 5.6.
3. Obtain baseline crack-growth rate data, i.e. da/dN versus ΔK (typically on a log-log scale) plots for the relevant material.
4. Determine the crack-growth curve using a crack-growth model such as the Paris equation if the spectrum is of constant amplitude, or a retardation model if the spectrum is of variable amplitude with overloads. It is standard accepted practice to start with a 0.05-inch flaw.
5. Once the crack growth curve is generated, regions for critical crack sizes can be identified using the results from previous steps, and appropriate inspection intervals established. Typical intervals are shown in Figure 5.7.

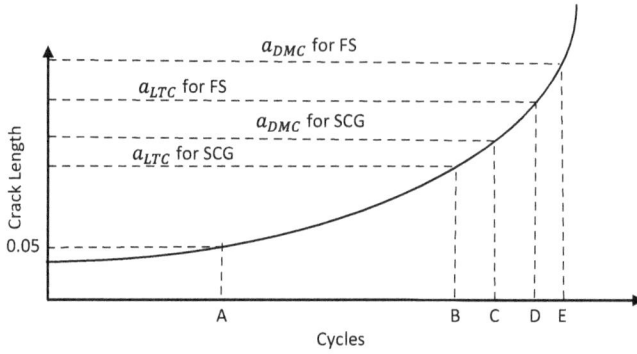

FIGURE 5.7   Crack growth curve showing typical DTA intervals.

Notes: The parameters are defined as follows. a_{LTC}-non-inspectable crack size: between $a = 0.05$ inch and $a = a_{LTC}$, no inspections are required; $a > a_{LTC}$: an inspection protocol is implemented; a_{DMC}-depot level inspectable crack size: time at which a basic inspection should be done. SCG = slow crack growth; FS = fail safe.

Example 5.2

A one-inch thick semi-infinite panel with an edge crack is subjected to a stress spectrum with an overload . The stress spectrum is represented by the stress ranges $\Delta\sigma = (17, 10, 22, 37, 52, 12)$ ksi. At the overload value, the crack length is 0.1 inch. The magnitude of the crack growth reduction factor, C_p, varies with crack length such that $C_p \approx 1.27(a/t)^{0.5}$ for $0.05 < a/t < 0.2$, where t is the panel thickness. If the crack growth rate for this material is given by $da/dN(\text{in/cycle}) = 0.61 \times 10^{-9}(\Delta K)^3$ with$\Delta\sigma$ and a in ksi and inches, respectively. Determine the reduced crack growth rate due to the overload.

From Eqn. (5.4)

$$\left(\frac{da}{dN}\right)_r = C_p f\left(\Delta K\right)$$

For an edge-cracked semi-infinite plate

$$K_I = 1.12\sigma\sqrt{\pi a} \Rightarrow \Delta K = 1.12\Delta\sigma\sqrt{\pi a}$$

Since the overload introduces crack retardation we apply the Wheeler model, therefore

$$\left(\frac{da}{dN}\right)_r = 1.27\left(a/t\right)^{0.5}\left(0.61\times10^{-9}\right)\left(\Delta K\right)^3$$

$$\left(\frac{da}{dN}\right)_r = 1.27\left(0.1\right)^{0.5}\left(0.61\times10^{-9}\right)\left(1.12\times\Delta\sigma\times\sqrt{\pi a}\right)^3$$

The overload will occur at the highest stress range, which is 52 ksi, therefore

$$\left(\frac{da}{dN}\right)_r = \left(0.24\times10^{-9}\right)\left(1.12\times52\times\sqrt{\pi\times0.1}\right)^3 = 8.52\times10^{-6}\text{in/cycle}$$

Example 5.3

The hoop stress history for an aircraft fuselage panel, $K_{Ic} = 56\,ksi\sqrt{in}$, 12 ft. x 6 ft., 0.1 in. thick, is known to be cyclical with a constant amplitude of 10 ksi and a minimum stress of 5 ksi. The baseline crack growth rate data can be modeled by $da/dN(in/cycle) = 0.39 \times 10^{-10}(\Delta K)^4$ with $\Delta\sigma$ and a in ksi and inches, respectively. Regulations indicate that depot level inspections should begin at a quarter lifetime and occur every 30,000 cycles subsequently. The panel should be taken out of service when $a = 0.5a_{crit}$ or at 200,000 cycles, whichever comes first. When should the depot level inspections start? How many depot level inspections will be needed for its useful service life?

SOLUTION

The stress amplitude, σ_a, is given by $\frac{\sigma_{max} - \sigma_{min}}{2}$, therefore $\sigma_{max} = 20 + 5 = 25\ ksi$.

The highest likelihood of failure would occur if a through crack is oriented perpendicular to the hoop stress. Therefore, this scenario can be modeled as a center through crack in an infinite panel. The governing equation is

$$K_I = \sigma\sqrt{\pi a}$$

The critical crack size is therefore found by

$$a_{crit} = \frac{1}{\pi}\left(\frac{K_{Ic}}{\sigma_{max}}\right)^2 = \frac{1}{\pi}\left(\frac{56}{25}\right)^2 = 1.6\,in$$

To determine the fatigue life, the initial crack size is assumed to be 0.05 inches.

$$N_f = \int dN = \int_{0.05}^{1.6} \frac{da}{0.39 \times 10^{-10}(\Delta K)^4}$$

$$= \int_{0.05}^{1.6} \frac{da}{0.39 \times 10^{-10}\left(20 \times \sqrt{\pi a}\right)^4}$$

$$= \int_{0.05}^{1.6} \frac{1.62 \times 10^{-10}}{a^2}\,da = 313,875\,cycles$$

Crack Growth Curve (a vs. N)

The crack growth curve can be generated by finding ΔN for incremental Δa.

At $a = 0.05$ inches, $\Delta a/\Delta N = 0.39 \times 10^{-10}\left(20\sqrt{\pi(0.05)}\right)^4 = 1.54 \times 10^{-7}\,in/cycle$.

Using increments of 0.01 for Δa, we obtain the following:

When $\Delta a = 0.01$, $\Delta N = \frac{0.01}{\left(1.54 \times 10^{-7}\right)} = 6.5 \times 10^4$ cycles

At $a = 0.06$ inches, $\Delta a/\Delta N = 0.39 \times 10^{-10}\left(20\sqrt{\pi(0.06)}\right)^4 = 2.22 \times 10^{-7}\,in/cycle$

When $\Delta a = 0.01$, $\Delta N = \dfrac{0.01}{\left(2.22 \times 10^{-7}\right)} = 4.5 \times 10^4$ cycles

Continued iterations will provide the data needed to plot crack size vs. number of cycles.

Inspections

Depot level inspections should begin at $0.25 \times 313{,}875 = 78{,}468$ cycles.

When $a = \dfrac{1}{2}a_{crit}$, the number of cycles, N, can be found by reading directly on the a vs. N curve or by performing another integration of the crack growth equation with an upper limit of $\dfrac{1}{2}a_{crit}$. Performing the latter yields

$$N = \int_{0.05}^{0.8} \frac{1.62 \times 10^4}{a^2}\, da = 303{,}750 \text{ cycles}$$

The number of cycles at $a = \dfrac{1}{2}a_{crit}$ exceeds 200,000; the panel should therefore be removed from service at 200,000 cycles. The number of inspections, N_{insp}, needed would be found from

$$N_{insp} = \frac{200{,}000 - 78{,}468}{30{,}000} = 4.$$

5.3   LEAK BEFORE BURST (LBB)

Since the 1950s, numerous investigations have been performed to assess the mechanical and structural behavior of pressurized components, such as the loading capacity and failure behavior of piping [5]. In the 1960s, Irwin [6] presented one of the first few cases associated with LBB. LBB usage is most prevalent in the nuclear industry; however, it finds application in aerospace, oil, transportation, and agriculture. The American Regulatory Authority, now called the United States Nuclear Regulatory Commission (USNRC) [7] provided the first guidance for the implementation, limitations, and acceptance of LBB in the 1980s. Other prominent LBB standards organizations include the American Petroleum Institute (API), the American Society of Mechanical Engineers (ASME), and the International Atomic Energy Agency (IAEA). Figure 5.8 shows a liquid oxygen (Lo2) pressure vessel at Launch Pad 39B at NASA's Kennedy Space Center, Florida-Pad 39B, undergoing a pressurization test after recent upgrades. The tank was designed to support NASA's Space Launch System (SLS). The application of FM methods to the design and testing of these pressure vessels is essential as service temperatures are in the −300 °F range and cryogenic oxygen can be pumped at rates as high as 85 pounds per second at pressures over 2,000 psi [8].

In order to prevent explosive failures in pressure vessels, an LBB design approach has been established where feasible. In this approach it is desirable that the crack propagates completely through the vessel wall thickness before it reaches critical length. The LBB concept seeks to demonstrate by deterministic FM that when a vessel is subjected to cyclical pressure, a crack would first grow through the wall,

FIGURE 5.8 Liquid oxygen (Lo2) tank at NASA's Kennedy Space Center undergoing a pressurization test.

Source: Image from NASA Kennedy Space Center blog [9].

resulting in a leak, before the pressure build up would be sufficient to cause a sudden explosion. The expectation is that the small "through wall" flaw would be detected by the plant's leakage monitoring systems long before the flaw could grow to an unstable size. Leakage exceeding some specified limit would require operator action or plant shutdown [10]. The cracks may start on the internal wall and propagate radially towards the outer wall or vice versa. Figure 5.9 shows a surface crack that has grown through the thickness to the inner wall.

5.3.1 ELLIPTICAL CRACK GROWTH BEHAVIOR

Consider an elliptical flaw with major axis length $2c$ and minor axis $2a$ embedded in a flat plate under tension, as shown in Figure 5.10. The elliptical crack is in the plane perpendicular to the loading direction and its major axis is parallel to the traction free surface. The stress intensity factor is at a maximum at the end of the minor axis, point A, and is at a minimum at the end of the major axis, point B. Irwin [6] showed that the crack-driving force would be greater in the radial direction than in the axial direction as long as the axial crack length was less than twice the cylinder thickness. Over

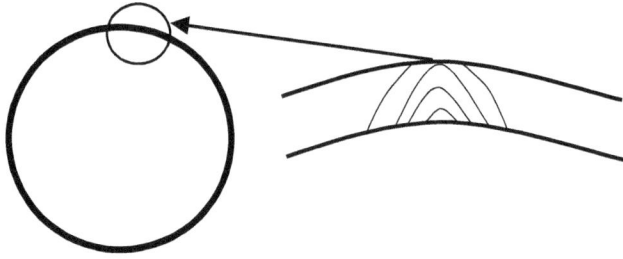

FIGURE 5.9   Crack growth in pressure vessel.

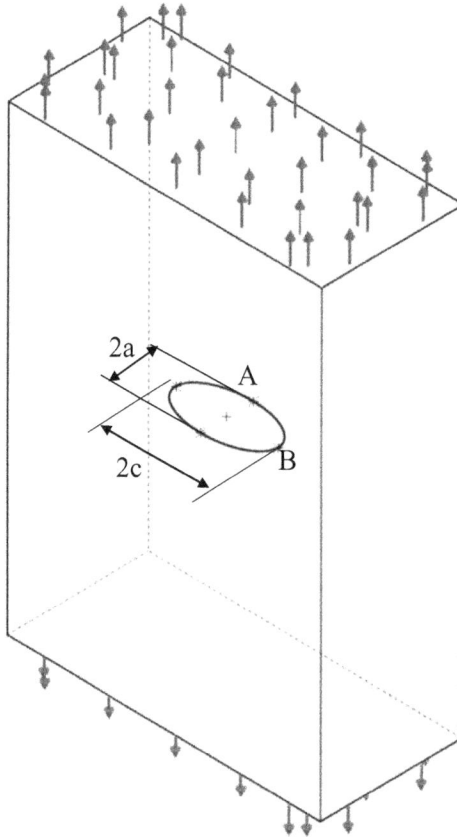

FIGURE 5.10   Embedded elliptical crack.

time the embedded elliptical crack will therefore grow such that the stress intensity values move towards a singular value. The faster growth in the minor axis direction occurs until a circular configuration is achieved (i.e. $a/2c = 0.5$). The crack will then continue to grow as a circular crack.

For the case of a semi-elliptical surface crack in a similar configuration, a true semicircle does not necessarily develop because the surface has an additional

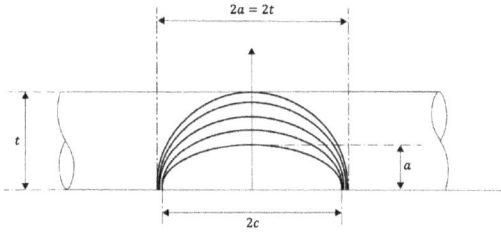

FIGURE 5.11 Semi-elliptical crack transitioning to a semi-circular crack as it grows through the thickness.

influence on K. The $a/2c$ ratio actually approaches 0.36; however, in a first-level LBB analysis, it is acceptable to approximate 0.5. LBB assumes any semi-elliptical surface crack growing through the thickness will first transition to a semi-circular configuration and then continue expanding until reaches the other side. At this point, the semi-circular crack breaks through the thickness and transitions to a through crack. The characteristic length of this through crack ($2a$) is taken as the diameter of the semi-circular crack at that point, ($2t$). This is shown in Figure 5.11. The time between breakthrough and the through crack achieving the length of the thickness is minimal, as only a small ligament of material would remain which is being subjected to hoop stresses as well as fluid pressure.

The stress intensity for this through crack is

$$K_I = \sigma\sqrt{\pi a} = \sigma\sqrt{\pi t} \tag{5.7}$$

Example 5.4

Calculate the maximum radius and wall thickness of a thin-walled spherical pressure vessel made from Ti-5Al-2.5Sn titanium alloy subjected to an in-service pressure of 475 kPa, so that it will leak from a surface crack before breaking. For this alloy $K_{Ic} = 51\,\text{MPa}\sqrt{m}$, $\sigma_{ys} = 320$ MPa.

Though the crack starts as a semi-elliptical surface crack, once it grows through the thickness it effectively becomes a through-the-thickness crack of length $2a$ where $a = t$, therefore

$$K_I = \sigma\sqrt{\pi t}$$

If $K_I < K_{Ic}$, the vessel will not fail via crack instability. Therefore, to find the maximum thickness let $K_I = K_{Ic}$. The vessel must also contain the pressure without yielding; therefore also let $\sigma = \sigma_{YS}$

$$51 = (320)\sqrt{\pi t}$$

$$t = 8.1\,\text{mm}$$

For a thin-walled spherical pressure vessel the stress is independent of direction and given by

$$\sigma = \frac{Pr}{2t}$$

The radius can now be found:

$$r = \frac{2t\sigma}{P} = \frac{2\left(8.1\times10^{-3}\right)\left(320\right)}{\left(475\times10^{-3}\right)} = 10.91\,\text{m}$$

Example 5.5

Nondestructive inspection techniques on a thin-walled cylindrical pressure vessel have revealed a small semicircular flaw (radius 0.3 cm) located at the inner surface and oriented normal to the hoop stress direction. The vessel is subjected to increasing pressure. Its wall is 2 cm thick, the material fracture toughness is 90 MPa$\sqrt{m}$, the yield strength is equal to 742 MPa. At a hoop stress of 325 MPa, would the vessel leak before it ruptured? The stress intensity factor for a semi-circular surface crack is given by $K_I = 1.12\left(\dfrac{2}{\pi}\right)\sigma\sqrt{\pi a}$.

SOLUTION

For leak-before-break conditions, critical crack length $a_c > t$, where t is the wall thickness.

If $a_c = t$ for a through crack, the stress intensity factor would be

$$K_I = 325\sqrt{\pi\left(2\times10^{-2}\right)}$$

$$K_I = 81.5\,\text{MPa}\sqrt{m}$$

$K_I = 81.5\,\text{MPa}\sqrt{m} < K_c = 90\,\text{MPa}\sqrt{m}$; therefore, the through crack has not yet reached its critical length. The vessel will leak before it bursts.

Alternatively, we can use the fracture toughness value to find the critical crack length, then compare the critical crack length to the wall thickness.

$$a_c = \frac{1}{\pi}\left(\frac{90}{325}\right)^2 = 2.4\,\text{cm}$$

$a_c = 2.4$ cm $> t = 2$ cm, so the crack length for leakage is less than its critical value. The vessel will therefore leak before it breaks.

Note: The stress intensity factor for a semi-circular flaw was not used since it must break through to become a through crack for leakage to occur.

Example 5.6

Consider the problem in Example 5.5, but that now the material obeys a crack growth law given by da/dN(m/cycle) $= 0.39 \times 10^{-8}(\Delta K)^4$, with $\Delta\sigma$ and a in MPa and m respectively. The radius of the closed cylinder is 5 m. If the pressure cycle range is 300 kPa twice per day, how long will it take before leakage starts assuming the semi-circular surface crack remained stable during growth through the thickness and the LBB criterion is satisfied?

SOLUTION

The pressure range is 300 kPa. The stress range may be found using

$$\sigma = \frac{Pr}{t} \Rightarrow \Delta\sigma = \frac{\Delta Pr}{t} = \frac{300 \times 10^{-3} \times 5}{2 \times 10^{-2}} = 75 \text{ MPa}$$

The stress intensity factor $K_I = 1.12 \left(\dfrac{2}{\pi}\right)\sigma\sqrt{\pi a} \Rightarrow \Delta K = 1.12\left(\dfrac{2}{\pi}\right)\Delta\sigma\sqrt{\pi a}$

Let the number of cycles required to grow through the thickness be denoted by N_t, then

$$N_t = \int dN = \int_{0.003}^{0.02} \frac{da}{0.39 \times 10^{-8}(\Delta K)^4}$$

$$= \int_{0.003}^{0.02} \frac{da}{0.39 \times 10^{-8}\left(1.12 \times \dfrac{2}{\pi} \times 75 \times \sqrt{\pi a}\right)^4}$$

$$= \int_{0.003}^{0.02} \frac{3.18}{a^2} da = 901 \, cycles$$

At two cycles per day, we obtain $\dfrac{901}{2}$ days or approximately 1.2 years.

Note: The stress intensity relationship for the semi-circular crack was used, as a through crack is assumed to manifest immediately once it breaks through the thickness, then leakage starts.

5.4   SUMMARY

The elementary aspects of DTA and LBB have been presented. A variable amplitude spectrum will substantially complicate DTA methods as the effect of overload, changing R-ratios, and crack closure become significant. The DTA method is also applied to built-up structures such as stiffened panels as opposed to single load path structures. The LBB approach does have limitations. Special adjustments must be made when weldments are present, or the material of concern is inhomogeneous (e.g. a layered pressure vessel or a composite). Before the implementation of LBB, the engineer should ensure that the appropriate measurement instruments are available to accurately detect the calculated leakage; otherwise, the assessment will be invalid. The engineer should also ensure that the leak can be properly managed prior to implementing LBB.

PROBLEMS

5.1. List and discuss the differences between a slow crack growth and a fail-safe structure for DTA qualification.

5.2. For a fail-safe structure, there is a requirement that the remaining structure at the time of a single load path failure must be capable of withstanding a minimum load. This minimum load is equal to the load that caused the load path failure plus an additional increment. Why might this be the case?

5.3. Would the fact that a structure has alternate load paths (local redundancy) in some locations necessarily qualify it as fail safe? Why/why not?

5.4. Use mathematical software to plot the residual strength curve for a circular crack in an infinite medium with plane strain fracture toughness $K_{Ic} = 95$ MPa$\sqrt{m}$. See Table 2.2

5.5. Use mathematical software to plot the residual strength curve for an edge cracked plate made from material with a plane strain fracture toughness $K_{Ic} = 23$ MPa$\sqrt{m}$. The plate width is 400 mm. The geometric factor for an edge cracked finite plate is given by $Y(a/W) = \dfrac{\left(1.12 + (a/W)\{2.91(a/W) - 0.64\}\right)}{\left(1 - 0.93(a/W)\right)}$.

5.6. A centrally cracked panel is subjected to a stress spectrum with an overload when $\Delta\sigma = 30$ ksi, at a crack half-length of 0.15 inches. The magnitude of the crack growth reduction factor, C_p, varies with crack length such that $C_p \approx 0.38a^{0.4}$ for $0.05 < a < 0.2$. If the crack growth rate for this material is given by $da/dN = 0.7 \times 10^{-11}(\Delta K)^2$, determine the reduced crack growth rate due to the overload.

5.7. A thick structural component is expected to undergo constant amplitude cyclical in-service loads with an R-Ratio of 0.2 and a stress amplitude of 17 ksi. The baseline crack growth rate data can be modeled by $da/dN = 0.21 \times 10^{-9}(\Delta K)^{2.5}$. Due to manufacturing imperfections, the most likely flaw is a circular crack in the center of the body perpendicular to the applied tensile loads. Regulations indicate that depot level inspections should begin at 1/5 lifetime and occur every 5,000 cycles subsequently. The component should be taken out of service when $a = 0.2a_{crit}$ or at 30,000 cycles, whichever comes first. When should the depot level inspections start? How many depot level inspections will be needed for its useful service life? The fracture toughness of the material is $K_{Ic} = 41$ ksi$\sqrt{in}$.

5.8. Calculate the maximum radius and wall thickness of a cylindrical pressure vessel made from 17-7PH steel subjected to an in-service pressure of 400 psi, so that it will leak from an initial semi-elliptical surface crack in a plane perpendicular to the hoop direction before breaking. For this material $K_{Ic} = 46$ ksi$\sqrt{in}$, $\sigma_{YS} = 120$ ksi.

5.9. A cylindrical pressure vessel made from 2024 Al-T3 ($K_{Ic} = 24$ ksi $\sqrt{in}$ $\sigma_{YS} = 24$ ksi is designed to operate in one of NASA's test facilities. The vessel must be designed such that the operating pressure should generate a hoop stress not to exceed $0.3\sigma_{YS}$. In order to meet NASA's Fitness for Service qualification a hydraulic proof test must be done where the hoop stress generated does not exceed $0.6\sigma_{YS}$. The cylinder has a wall thickness of two inches and a diameter of six feet. A semi-circular surface flaw 0.5 inches deep, perpendicular to the hoop stress, is discovered just before the proof test. Will it survive the proof test? Will it leak before it breaks?

REFERENCES

[1] A. Fajri, A. Prabowo, N. Muhayat, D.F. Smaradhana and A. Bahatmaka, "Fatigue Analysis of Engineering Structures: State of Development and Achievement," *Procedia Structural Integrity*, vol. 33, pp. 19–26, 2021.

[2] O. Gunes, "Failure modes in structural applications of fiber-reinforced polymer (FRP) composites and their prevention," in *Developments in Fiber-Reinforced Polymer (FRP) Composites for Civil Engineering*, Sawston, Woodhead Publishing Ltd., 2013, pp. 509–525.

[3] P. Miedlar, A. Berens, A. Gunderson and J.P. Gallagher, USAF Damage Tolerant Design Handbook: Guidelines for the Analysis and Design of Damage Tolerant Aircraft Structures, Dayton, OH: University of Dayton Research Institute, 2002.

[4] O. Wheeler, "Spectrum Loading and Crack Growth," *Trans. ASME J. Basic Eng.*, vol. 94, p. 181, 1972.

[5] R. Bourgaa, P. Mooreb, Y.-J. Janinb, B. Wanga and J. Sharples, "Leak-before-break: Global Perspectives and Procedures," *International Journal of Pressure Vessels and Piping*, Vols. 129–130, pp. 43–49, 2015.

[6] G. Irwin, "Materials for Missiles and Spacecraft," in *Fracture of Pressure Vessels*, Mcgraw Hill, New York City, 1963, pp. 204–229.

[7] United States Nuclear Regulatory Commission (USNRC), "Standard Review Plan 3.6.3 Rev. 1 Leak Before Break Evaluation Procedures. No. 301 NUREG-800," United States Nuclear Regulatory Commission (USNRC), Rockville, MD, 2007.

[8] S.H.S. Forth, P. Gregory, B. Mason, J. Thompson and E. Hoffman, "Damage Tolerance Analysis of a Pressurized," NASA Langley Research Center, NASA/TM-2006-214274, Hampton VA, 2006.

[9] L. Herridge, "Chilling Out During Liquid Oxygen Tank Test," NASA Keneddy Space Center Blog (https://blogs.nasa.gov), Hampton VA, 2018.

[10] International Atomic Energy Agency IAEA, "Applicability of the Leak Before Break Concept," INIS Clearing House, International Atomic Energy Agency, Vienna, Austria, 1993.

6 Experimental Methods

6.1 MEASUREMENT OF FRACTURE TOUGHNESS

Previously, analytical relationships based on the stress intensity factor K for linear elastic fracture mechanics (LEFM) and the J integral and crack tip opening displacement (CTOD) for elastic plastic fracture mechanics (EPFM) were discussed. However, in order to successfully predict failure when designing engineering structures and components, the critical values of these parameters need to be assessed through experimental testing [1]. The measurement of the fracture toughness provides an indication of a material's resistance to crack extension. Testing methods can yield a single value of fracture toughness like K_{IC}, or the toughness may be plotted against crack extension to produce a resistance curve.

This chapter describes various experimental techniques for the measurement of fracture toughness according to ASTM testing standards, including procedures for K_{IC}, J_{IC}, and CTOD as well as K-R curve testing and ductile-to-brittle transition temperature testing for metals. Although the most recent standards are discussed here, they continue to evolve as our experience and as technology improves. As such, it is left to the reader to consult the relevant standards for their application.

6.1.1 SPECIMEN TYPE

There are five types of specimens allowed in the ASTM standard which include the compact specimen, the arc-shaped specimen, single-edge-notched bend (SE(B)) geometry, the disk specimen, and the middle tension (MT) specimen. There are three important dimensions for these configurations: the specimen thickness (B), width (W), and crack length (a). The compact and SE(B) are the two most popular fracture toughness tests. Figure 6.1 provides an illustration of these specimens with characteristic dimensions (B, W, a).

Specimen size can be scaled geometrically. ASTM standard sizes include 1/2T, 1T, 2T, and 4T where the numerical value refers to the specimen thickness in inches. A standard 1T compact specimen, for example, has dimensions B = 1 in (25.4 mm) and W = 2 in (50.8 mm).

6.1.2 SPECIMEN ORIENTATION

The orientation of the specimen can have a pronounced effect on the fracture toughness values as engineering materials are rarely homogeneous and isotropic. This sensitivity to orientation is tied to a material's microstructure, which may contain planes

(a) Single Edge Notched Bend (SE(B)) Specimen

(b) Middle Tension (MT) Specimen

(c) Compact Specimen

(d) Disk Shaped Compact Specimen

(e) Arc-shaped Specimen

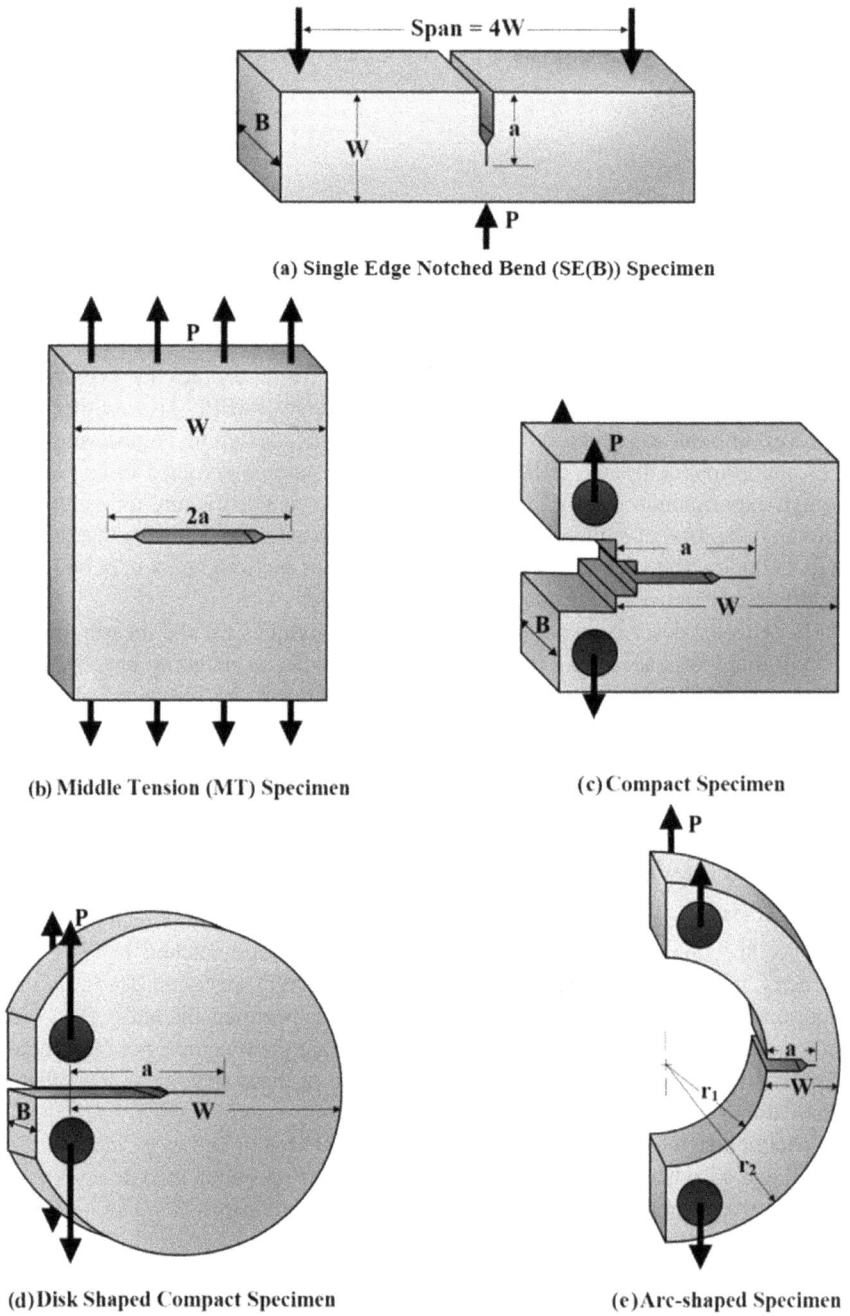

FIGURE 6.1 ASTM standard fracture test specimens.

of weakness that allow cracks to easily propagate. As a result, the ASTM testing standards for fracture toughness measurement require that the orientation also be specified when reporting test results.

Figure 6.2 shows the orientation including the two-letter notation adopted by ASTM [2]. For rectangular section specimens produced from a rolled plate or forging the orientations are defined using the letters L, T, and S, where

L = direction of principal deformation (maximum grain flow)
T = direction of least deformation
S = third orthogonal direction.

For cylindrical bars and tube specimen, orientations are defined as shown in Figure 6.3. Here the letters used are L, C, and R, where L = axial direction; R = radial direction; C = circumferential or tangential direction.

Although determining the fracture toughness of a material should involve tests in several orientations, this is not always feasible. In selecting the appropriate specimen orientation, one should not only consider the constraints that the material is under but also the reason for using the test. In the case of general material characterization, a low toughness orientation such as S-L or T-L that allows crack propagation in the rolling or forging direction should be chosen. When attempting to estimate toughness in a flawed structure, the orientation that most closely matches that of the actual flaw should be considered.

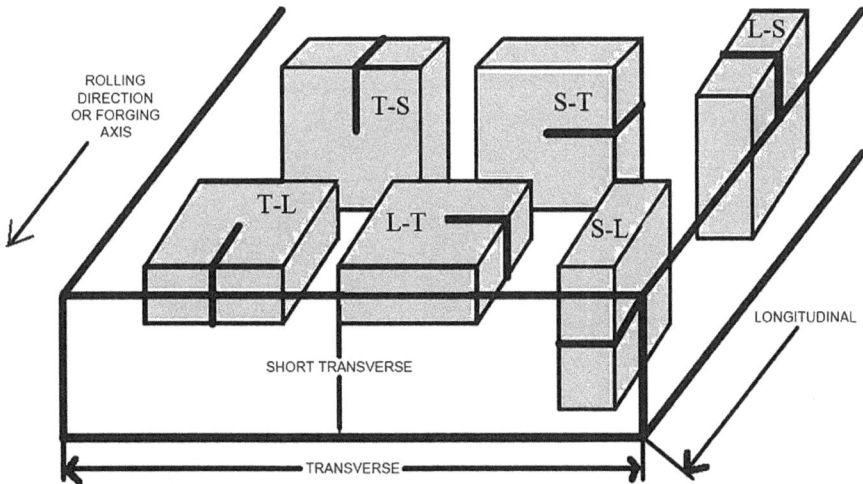

FIGURE 6.2 ASTM notation and orientation for rectangular section specimens obtained from rolled plate and forgings.

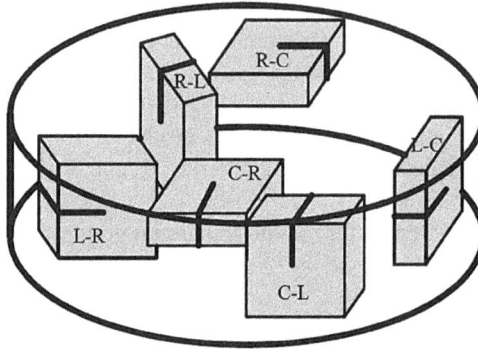

FIGURE 6.3 ASTM notation and orientation for specimens obtained from cylindrical bars and tubes.

6.1.3 MEASUREMENT APPARATUS

The facture toughness test requires, at a minimum, measurement of both applied load and crack mouth opening displacement (CMOD). Load measurements are easily accomplished with tensile testing equipment containing a load cell. For displacements measurements, ASTM E399-20a [3] recommends the use of a clip gauge as shown in Figure 6.4. The arms of the gauge attach to the mouth of the crack by making contact with sharp knife edges. These ensure free rotation of the arms and can either be machined or attached to the specimen as illustrated. Four strain gauges are bonded to the arms to form a Wheatstone Bridge circuit. As the test is conducted the clip gauge arms deflect, resulting in a voltage change across the bridge, which varies linearly with displacement. It is recommended that triplicate tests be performed at a minimum for each material condition.

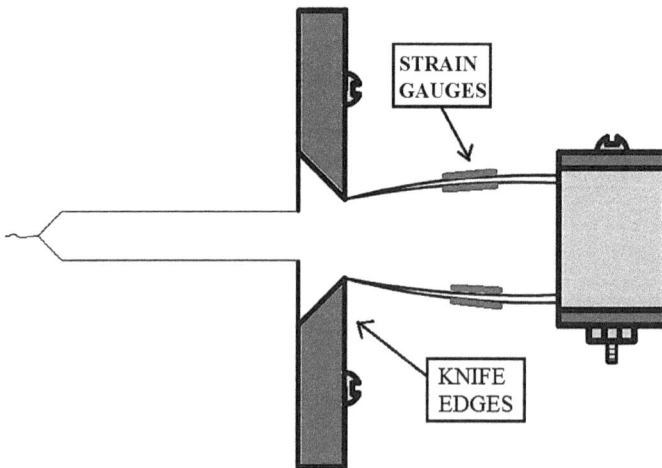

FIGURE 6.4 Crack mouth opening displacement measurement using a clip gauge.

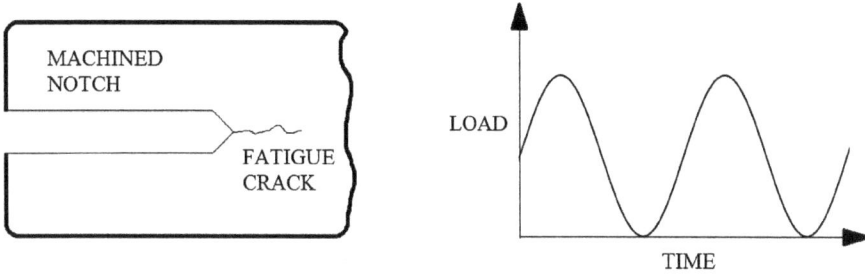

FIGURE 6.5 Fatigue pre-cracking of fracture toughness specimen.

6.1.4 SPECIMEN PREPARATION

Fracture toughness tests require the presence of a sufficiently sharp crack. This is achieved by subjecting the test specimens to cyclic loading as shown in Figure 6.5. The crack is introduced at the tip of a machined notch and grown to the desired length by controlling the applied cyclic load.

According to ASTM E399, the pre-crack size in terms of the length/width ratio (a/W) for compact specimens should satisfy

$$0.45 \leq a/W \leq 0.55 \tag{6.1a}$$

The presence of the initial fatigue crack should not adversely influence the fracture toughness measurement of the material being tested. This requires the establishment of a sharp-crack condition at the tip of the fatigue crack to ensure predominantly linear-elastic, plane strain conditions. In order for a K_{IC} test to be considered valid, ASTM E399-20a requires that the specimen ligament size ($W - a$) satisfies the following:

$$\left(W - a\right) \geq 2.5 \left(\frac{K_{IC}}{\sigma_{YS}}\right)^2 \tag{6.1b}$$

It is recommended that the user perform a preliminary check to determine and verify the appropriate specimen dimensions by estimating the anticipated K_{IC} value using data from similar materials.

6.1.5 K TESTING

The critical value of the Mode I stress intensity factor can be used as the fracture parameter when a material exhibits linear-elastic behavior. The assumption is that the plastic zone is small compared to the dimensions of the specimen. The test method described here is based on the ASTM E399-20a standard for the measurement of crack-extension resistance at the onset (2% or less) of crack extension for plane-strain loading conditions.

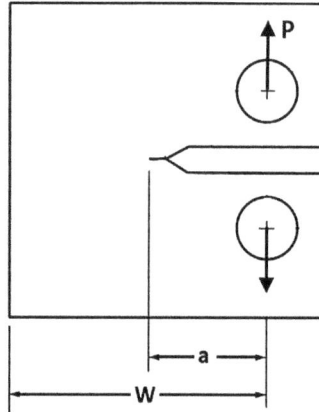

FIGURE 6.6 Compact test specimen in tension.

The standard compact specimen configuration is a single-edge-notched one with a pre-crack loaded in tension as shown in Figure 6.6. Clevis pins are used to minimize friction effects during specimen loading. Typically, K_{IC} specimens are fabricated such that the width W is twice the thickness B). The loading rate for a ($W/B = 2$) specimen is between 0.33 and 1.67 kN/s (4.5 to 22.5 klbf/min).

Concept Challenge 6.1

How does the use of Clevis pins reduce the effect of friction?

6.1.6 INTERPRETATION OF RESULTS

During the test, load and displacement are monitored continuously as the specimen is loaded to failure. Three types of load versus CMOD curves are shown in Figure 6.7. The critical load P_Q is required for computing the fracture toughness and can be defined in various ways. The first step is to construct a 5% secant line with a slope equal to 95% of the initial elastic loading curve slope, to determine the P_5 value. This line corresponds to a roughly 2.0% apparent crack extension. With Type I behavior, the load–displacement curve is relatively smooth with only a slight deviation from linearity before reaching the maximum load P_{max}. In the case of a Type I curve, $P_Q = P_5$. For a Type II curve, a small amount of unstable crack growth known as "pop in" occurs before the 5% deviation from linearity occurs. This pop-in value is used as P_Q in the Type II case. In Type III curves, where the maximum force preceding P_5 is larger than P_5, $P_Q = P_{max}$.

It is also specified by E399, that the specimen thickness should be chosen so that

$$\frac{P_{max}}{P_Q} \leq 1.10 \tag{6.1c}$$

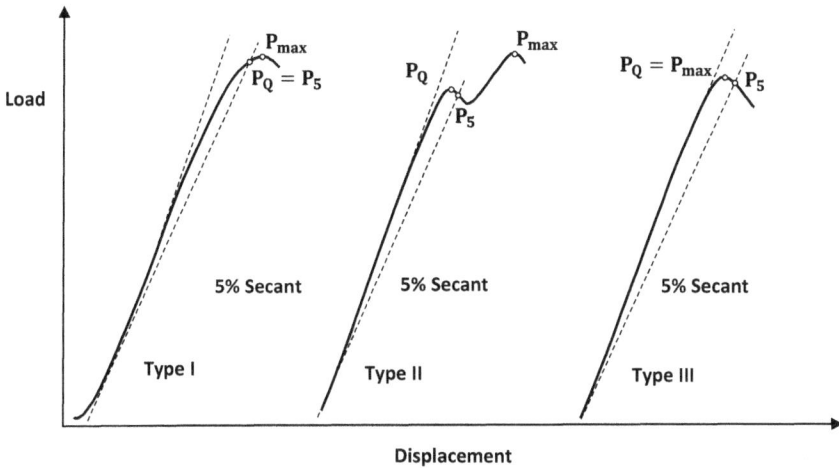

FIGURE 6.7   Three types of load versus CMOD curves in K_{IC} testing.

where P_{max} is the maximum force the specimen can sustain. For the compact speci-men, once P_Q is determined, a provisional fracture toughness K_Q is calculated using

$$K_Q = \frac{P_Q}{B\sqrt{W}} \cdot f\left(\frac{a}{W}\right) \tag{6.2}$$

where B is the specimen thickness, a is the crack size, and W is the specimen width specified in meters or inches. The dimensionless function $f(a/W)$ is given by ASTM E 399 as:

$$f\left(\frac{a}{W}\right) = \frac{\left(2+\dfrac{a}{W}\right)\left[0.866+4.64\left(\dfrac{a}{W}\right)-13.32\left(\dfrac{a}{W}\right)^2+14.72\left(\dfrac{a}{W}\right)^3-5.6\left(\dfrac{a}{W}\right)^4\right]}{\left(1-\dfrac{a}{W}\right)^{3/2}} \tag{6.3}$$

The fracture toughness value for K_Q computed using Eqn. (6.2) can be used as a valid K_{IC} result as long as all validity requirements in the standard (including those of Eqn. 6.1) are met.

Example 6.1

A compact specimen with a thickness of 25.4 mm and width of 50.8 mm, containing a 26.9 mm crack, is used in a fracture toughness test performed according to the E399 testing standard. Given $P_Q = 41.9$ kN, $P_{max} = 45.8$ kN, and $\sigma_{YS} = 750$ MPa, calculate K_Q and determine if it can be used as a fracture toughness measurement for this material.

Using Eqn. (6.3) for $f\left(\dfrac{a}{W}\right)$

$$f\left(\dfrac{a}{W}\right) = \dfrac{\left(2 + \dfrac{0.0269}{0.0508}\right)\left[0.866 + 4.64\left(\dfrac{0.0269}{0.0508}\right) - 13.32\left(\dfrac{0.0269}{0.0508}\right)^2 + 14.72\left(\dfrac{0.0269}{0.0508}\right)^3 - 5.6\left(\dfrac{0.0269}{0.0508}\right)^4\right]}{\left(1 - \dfrac{0.0269}{0.0508}\right)^{3/2}}$$

$$= 10.45$$

and

$$K_Q = \dfrac{(0.0419\,\text{MN})(10.45)}{(0.0254\,\text{m})\sqrt{0.0508\,\text{m}}} = 76.5\,\text{MPa}\sqrt{\text{m}}$$

In order for $K_Q = K_{IC}$, all validity requirements must be met.
(i)   From Eqn. (6.1a)

$$\dfrac{a}{W} = 0.53$$

which lies between the allowable limits of 0.45 and 0.55. So it is a pass.

(ii)   From Eqn. (6.1b)

$$(W - a) = 23.9\,\text{mm} \geq 2.5\left(\dfrac{K_{IC}}{\sigma_{YS}}\right)^2$$

$$2.5\left(\dfrac{76.5\,\text{MPa}\sqrt{\text{m}}}{750\,\text{MPa}}\right)^2 = 0.026\,\text{m} = 26.0\,\text{mm}$$

The specimen does not meet this requirement. So it is a fail.

(iii)   Based on Eqn. (6.1c),

$$\dfrac{P_{max}}{P_Q} = \dfrac{0.0458\,\text{MPa}}{0.0419\,\text{MPa}} = 1.09 < 1.10.\ \text{So it is a pass.}$$

Since *all* criteria for E 399 have not been met, $K_Q \neq K_{IC}$.

Concept Challenge 6.2

Toughness testing is performed according to the ASTM 399-20b standard on the compact specimens of a commercially available aluminum alloy. Is the fracture toughness obtained from the test useful in predicting the failure of a metal sheet of the same alloy containing a through crack? Why or why not?

6.1.7 J TESTING

The current ASTM E1820-20b [4] covers procedures and guidelines for determining fracture toughness or metallic materials using J-integral testing. K testing was included in earlier versions of the standard, but has since been removed to avoid duplication and is now handled under ASTM E399. The recommended test specimens are SE(B) compact, and disk-shaped compact. All specimens contain fatigue pre-cracks as previously described.

On loading the test specimen, either or both of the following may be induced: (1) unstable crack extension and significant pop-in referred to as "fracture instability" or (2) stable crack extension also known as "stable tearing." The fracture instability case results in a single point, fracture toughness value determined at the point of instability, while stable tearing allows a continuous fracture toughness versus crack-extension relationship or R-curve to be produced. If fracture instability interrupts the stable tearing event, an R-curve up to the point of instability results. E1820 describes two procedures for measuring crack extension for J testing: the basic and the resistance curve procedure which will be described in this section.

6.1.7.1 The Basic Method

This procedure involves loading the specimen to a desired displacement level and determining the amount of crack extension resulting from the loading. Since crack growth is not monitored during the process, the use of crack extension measurement equipment is not needed but multiple specimens are required. After the specimen is unloaded, the crack growth is physically marked using either heat tinting or fatigue cracking. The specimen is then carefully broken open to allow the crack extension to be measured. According to the ASTM standard for the basic procedure, a simplified approach to computing J is given by

$$J = J_{el} + J_{pl} \tag{6.4}$$

where J_{el} and J_{pl} are the elastic and plastic components of J respectively. J_{el} is given by

$$J_{el} = \frac{K^2 \left(1 - \upsilon^2\right)}{E} \tag{6.5}$$

K is determined based on load and crack size using Eqn. (6.2); J_{pl} can be estimated using

$$J_{pl} = \frac{\eta A_{pl}}{B_N b_O} \tag{6.6}$$

where A_{pl} is the plastic area under the load-displacement curve (Figure 6.8), B_N is the net specimen thickness ($B_N = B$ for specimens with no side grooves), and b_O is

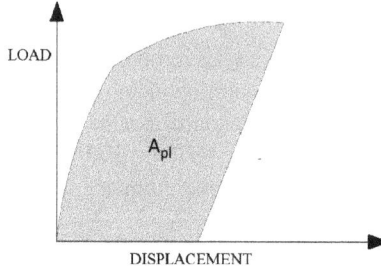

FIGURE 6.8   Load-displacement graph showing plastic energy absorbed by specimen during *J* testing.

the initial ligament length. The dimensionless constant η, for compact specimens, is given by

$$\eta = 2 + 0.522\, b_0 / W \tag{6.7}$$

whereas for SE(B) specimens $\eta = 2$. The values obtained using Eqn. (6.6) and (6.7) are based on the initial crack length and do not correct for crack growth when computing *J*.

An *R* curve is generated using the computed *J* values and the crack extension; (Δa) is obtained from physical test measurements and is given by

$$\Delta a = a_p - a_0 \tag{6.8}$$

where a_0 is the original crack size and a_p is the final physical crack size.

E 1820-20b outlines a procedure for determining J_Q from this R curve as illustrated in Figure 6.9. A construction line is first drawn using

$$J = 2\sigma_Y \Delta a \tag{6.9}$$

where σ_Y is the flow stress. An exclusion line parallel to the construction line is drawn to intersect the abscissa at 0.15 mm and 1.5 mm. A horizontal exclusion line is drawn with a limit value for *J* defined by

$$J_{limit} = \frac{b_0 \sigma_Y}{6.5} \tag{6.10}$$

A power law expression is used for fitting all data that fall within this region and defined by

$$J = C_1 \left(\Delta a \right)^{C_2} \tag{6.11}$$

Finally, a 0.2 mm offset line from the construction line is drawn to intersect the power law line and define J_Q. Assuming the size requirements given by

$$B, b_0 > \frac{10 J_Q}{\sigma_Y} \tag{6.12}$$

are satisfied along with all other validity criteria, then $J_Q = J_{IC}$.

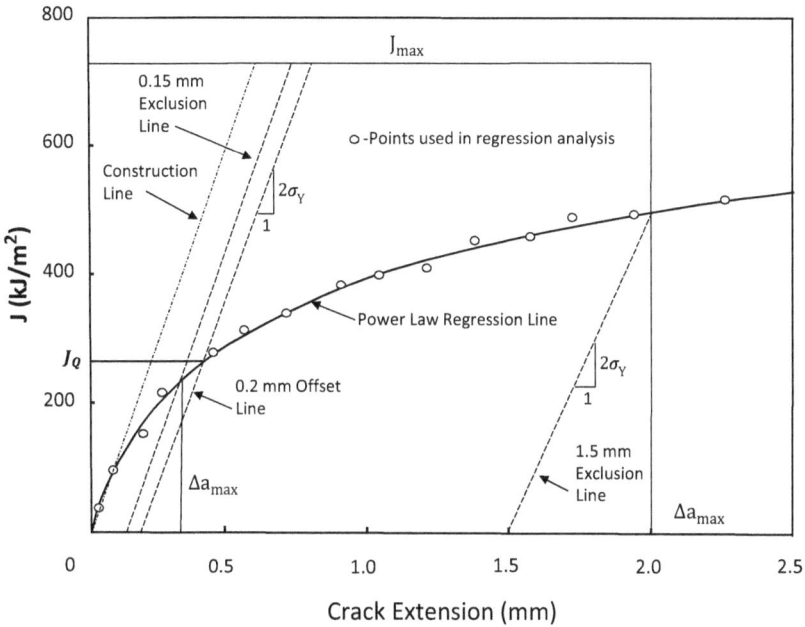

FIGURE 6.9 *J-R* curve used for finding J_Q.

Source: Reprinted, with permission, from ASTM E1820-20b Standard Test Method for Measurement of Fracture Toughness; copyright ASTM International, 100 Barr Harbor Drive, West Conshohocken, PA 19428.

6.1.7.2 Resistance Curve Method

For the resistance curve procedure, the crack growth is monitored during the test with the aid of crack extension measuring equipment. These data can be used to develop the characteristic *J-R* curve which consists of a plot of J versus crack extension as illustrated in Figure 6.10. For single specimen testing, the most common method for monitoring crack growth is the unloading compliance method.

The procedure requires the crack length to be computed at regular intervals during the test by allowing the specimens to be partially unloaded and measuring the compliance as shown in Figure 6.11. ASTM specifies the appropriate compliance equations for bend and compact specimens.

In order to avoid effects such as crack tunneling and shear lip formation, specimens should be side-grooved. The J integral is computed incrementally since the crack length changes continuously during testing. For compact specimens, at the *i*th measuring point, J is defined as

$$J_{(i)} = \frac{\left(K_{(i)}\right)^2 \left(1-v^2\right)}{E} + J_{pl(i)} \tag{6.13}$$

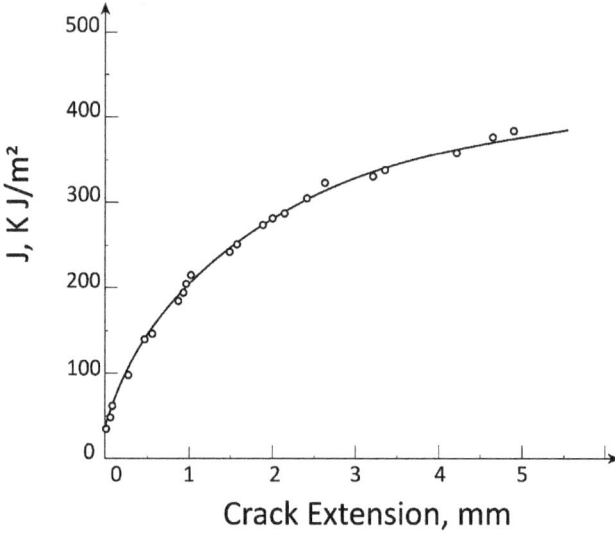

FIGURE 6.10   Sketch of a typical *J-R* curve.

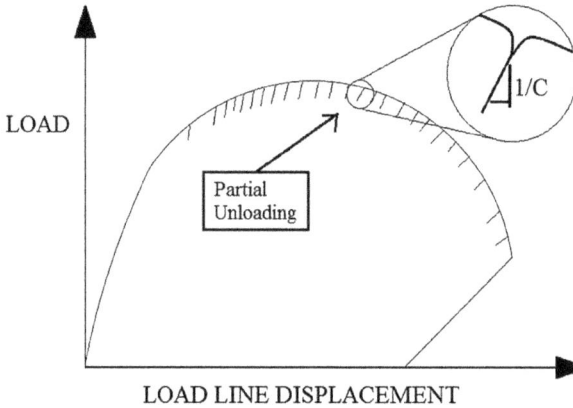

FIGURE 6.11   Crack growth monitoring for *J* testing using the unloading compliance method.

$K_{(i)}$ is computed from Eqn. (6.2) with $a = a_0$ and

$$J_{pl(i)} = \left[J_{pl(i-1)} + \left(\frac{\eta_{pl(i-1)}}{b_{(i-1)}} \right) \frac{A_{pl(i)} - A_{pl(i-1)}}{B_N} \right] \left[1 - \gamma_{(i-1)} \left(\frac{a_i - a_{(i-1)}}{b_{(i-1)}} \right) \right] \qquad (6.14)$$

where $\eta_{pl(i-1)} = 2.0 + 0.522\, b_i / W$ and $\gamma_{(i-1)} = 1.0 + 0.76\, b_i / W$.

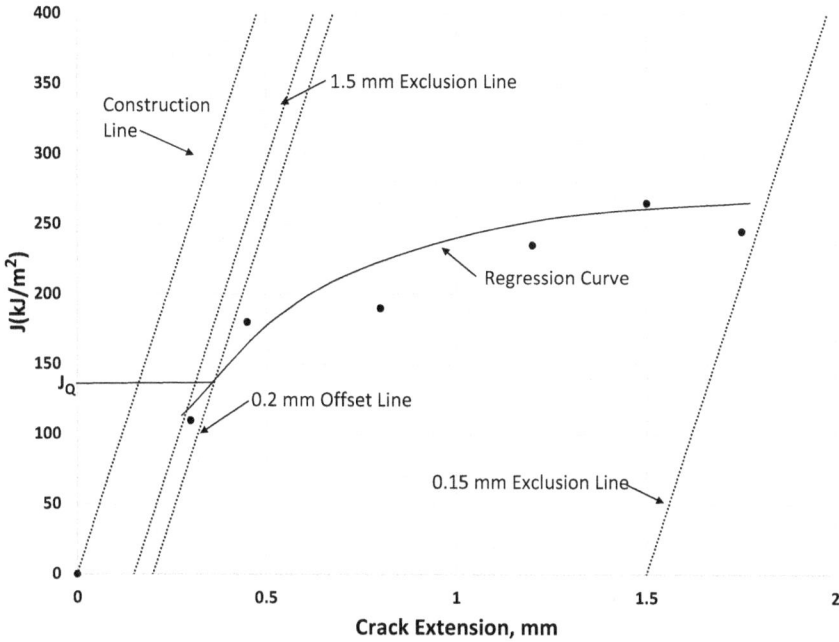

FIGURE 6.12 *J-R* curve plot.

The *J-R* curve is generated using the computed *J* values and crack extension Δa_i determined at each a_i as described in the standard. The maximum *J* capacity and crack extension for a specimen is given by

$$J_{max} = \frac{b_0 \sigma_Y}{10} \tag{6.15}$$

and

$$\Delta a_{max} \leq 0.25 b_0 \tag{6.16}$$

The curve fitting approach previously described can be used to obtain J_Q. Assuming all ASTM criteria are valid including size criteria, $J_Q = J_{IC}$.

Example 6.2

Fracture tests on multiple specimens were conducted according to the E 1820 J testing procedure. Values for J and crack extension were measured and used to generate a *J-R* curve. The specimen thickness was 25 mm with an initial ligament length of 24 mm. For the specimen, $\sigma_{YS} = 380$ MPa and $\sigma_{TS} = 460$ MPa. Using the tabulated data, construct the *R* curve and determine J_Q and if possible J_{IC}.

Specimen #	J, kJ/m²	Crack Extension, mm
1	110	0.3
2	180	0.45
3	190	0.8
4	235	1.2
5	265	1.5
6	245	1.75

Using the data provided, the R curve is plotted as in Figure 6.12.

Applying the ASTM E 1820 standard procedure, $J_Q = 136$ kJ/m². According to the standard, the following criteria need to be satisfied for assessing J_{IC}:

(i)   From Eqn. (6.12)

$$b_0 = 24\,mm > \frac{10J_Q}{\sigma_Y}$$

$$\frac{10\left(0.155 MJ/m^2\right)}{(380+460)\,MPa/2} = 0.0037\,m = 3.7\,mm. \text{ Pass.}$$

(ii)   Using Eqn. (6.15)

$$J_{max} = \frac{b_0\sigma_Y}{10} = \frac{0.024\,m\left(420\,MPa\right)}{10} = 1.008\,MJ/m^2$$

J_{max} is larger than all measured values of J. **Pass.**
Since all ASME E 1820 validity criteria have been met, $J_Q = J_{IC} = 136$ kJ/m².

Concept Challenge 6.3

Which measurement is a more conservative estimate of fracture toughness: J_{IC} or K_{IC}? Explain the reason for your choice.

6.1.8   CTOD TESTING

The ASTM E1820-20b standard also provides test procedures for determining the fracture toughness of a material based on the CTOD parameter, δ. As with J-testing, E 1820-20b includes a basic and resistance curve procedure when performing CTOD testing. The basic procedure which can be used to determine fracture toughness at the onset of ductile tearing does not consider stable crack growth in the analysis. For a SE(B) specimen, the CTOD is estimated using

$$\delta = \frac{J}{m\sigma_Y} \tag{6.17}$$

where J is defined using Eqn. (6.4) and m is given by

$$m = A_0 - A_1\left(\frac{\sigma_{YS}}{\sigma_{TS}}\right) + A_2\left(\frac{\sigma_{YS}}{\sigma_{TS}}\right)^2 - A_3\left(\frac{\sigma_{YS}}{\sigma_{TS}}\right)^3 \qquad (6.18)$$

where $A_0 = 3.18 - 0.22(a_0/W)$, $A_1 = 4.32 - 2.23(a_0/W)$, $A_2 = 4.44 - 2.29(a_0/W)$, and $A_3 = 2.05 - 1.06(a_0/W)$.

Measurements are made for crack extension Δa with multiple specimens and used to produce a δ-R curve. ASTM E 1820-20b outlines a similar procedure for finding δ_Q using this resistance curve and the power law curve fitting procedure, as previously described when determining J_Q. If all criteria specified by the ASTM standard are met, including the size requirement, given by

$$b_0 \geq 10m\delta_Q \qquad (6.19)$$

where m is defined by Eqn. (6.18), then $\delta_Q = \delta_{IC}$, which qualifies as a valid measure of fracture toughness.

The resistance curve method can also be used to produce a δ-R curve from a single specimen, using the E 1820-20b standard. For a SE(B) specimen, δ is estimated using

$$\delta_i = \frac{J_i}{m_i\sigma_Y} \qquad (6.20)$$

where J_i is computed using Eqn. (6.13) and m is defined by Eqn. (6.18).

The δ-R curve is generated from the computed δ_i and measured crack extension Δa_i values determined at each of the ith measuring points as illustrated in Figure 6.13.

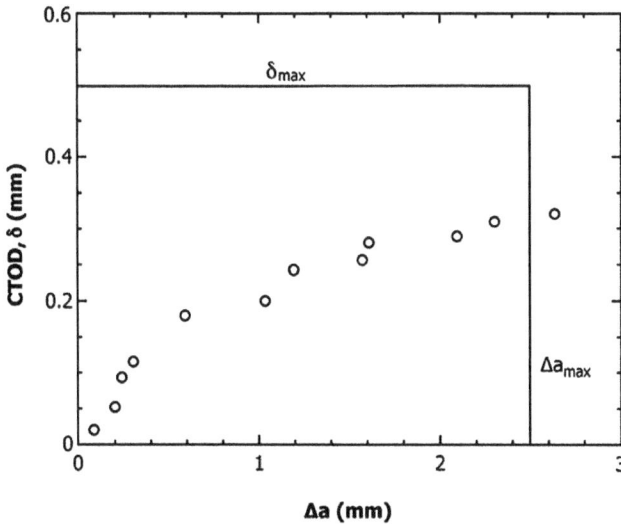

FIGURE 6.13 Sketch of crack tip opening displacement versus crack extension.

Specimens must satisfy all validity criteria including the maximum δ and crack extension capacity given by

$$\delta_{max} = \frac{b_0}{10m} \tag{6.21}$$

and

$$\Delta a_{max} = 0.25 b_0 \tag{6.22}$$

A similar power law curve fitting procedure previously discussed can be used for finding δ_Q which can be taken as δ_{IC}, provided the necessary criteria are met. E 1820-20b also provides procedures for cases in which specimens experience fracture instability before and after stable tearing. Although these will not be discussed here, the reader is encouraged to consult the standard for further details.

Example 6.3

An SE(B) specimen with $B = W = 25.4$ mm is used to perform a CTOD test. The initial crack depth a_0 is 11.7 mm and failure occurred with no prior stable crack growth. The computed J is 0.274 MJ/m². Compute the CTOD for this test, given $\sigma_{YS} = 345$ MPa, $\sigma_{TS} = 483$ MPa.

In order to find m, we must first solve for $A_0, A_1, A_2,$ and A_3:

$$(a_0/W) = (11.7\,\text{mm}/25.4\,\text{mm}) = 0.461$$

$$A_0 = 3.18 - 0.22(a_0/W) = 3.078$$

$$A_1 = 4.32 - 2.23(a_0/W) = 3.293$$

$$A_2 = 4.44 - 2.29(a_0/W) = 3.385$$

$$A_3 = 2.05 - 1.06(a_0/W) = 1.562$$

Using Eqn. (6.18):

$$m = 3.078 - 3.293\left(\frac{345}{483}\right) + 3.385\left(\frac{345}{483}\right)^2 - 1.562\left(\frac{345}{483}\right)^3 = 1.884$$

From Eqn (6.17), δ is given as

$$\delta = \frac{J}{m\sigma_Y} = \frac{0.274\,\text{MJ/m}^2}{1.884\left((345\,\text{MPa} + 483\,\text{MPa})/2\right)} = 0.00035\,\text{m} = 0.35\,\text{mm}$$

6.2 IMPACT TESTING

6.2.1 CHARPY AND IZOD TESTING

Impact testing is normally performed to better understand the effects of dynamic loading on a material. These tests provide a single value of the energy absorbed during impact for a single specimen. The most common of these are the Charpy and Izod impact test and the drop weight test. These tests can provide a low cost, convenient, and somewhat reliable indication of material toughness when compared to fracture mechanics testing. However, they lack the predictive capabilities and rigor compared to the fracture mechanics methods. One key difference is that the Charpy and Izod specimens contain a blunt notch while fracture mechanics test specimens have a sharp fatigue crack. Fracture tests are also conducted under quasi-static conditions while the Charpy and Izod specimens experience impact loading.

The ASTM E 23-18 [5] standard outlines procedures for both the Charpy and Izod impact testing of metallic materials. These tests involve measuring the energy absorbed by notched bar specimens from pendulum impact. In the case of the Charpy test, the specimen is a simple beam in three-point bending while for the Izod the specimen is a cantilever beam fixed at one end and impacted at the free end as shown in Figure 6.14.

The standard Charpy specimen has a 10 × 10 mm cross-section and is 55 mm long. Izod specimens have the same cross-section but are 75 mm long. Their relatively small size minimizes material consumption during testing. Figure 6.15 illustrates the pendulum apparatus typically used in these tests.

6.2.2 INTERPRETATION OF RESULTS

The pendulum is released from a known height y_1 and swings freely, colliding with the specimen to reach a height y_2. The energy absorbed E_F is given by

$$E_F = W\left(y_1 - y_2\right) - E_L \qquad (6.23)$$

where
   W is the weight of the hammer and
   E_L represents energy losses due to friction and drag.

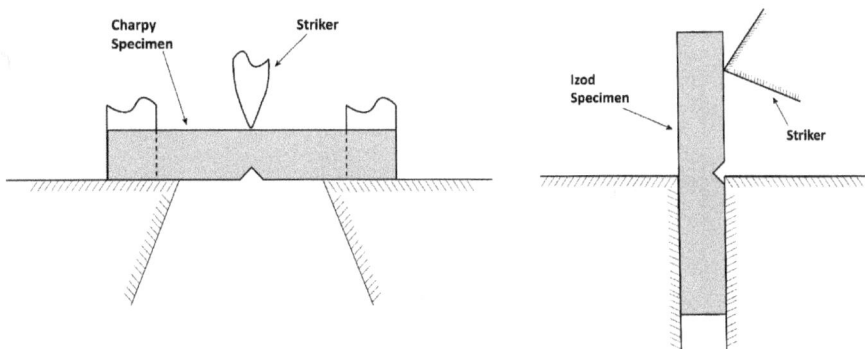

FIGURE 6.14   Impact testing of Charpy and Izod specimens.

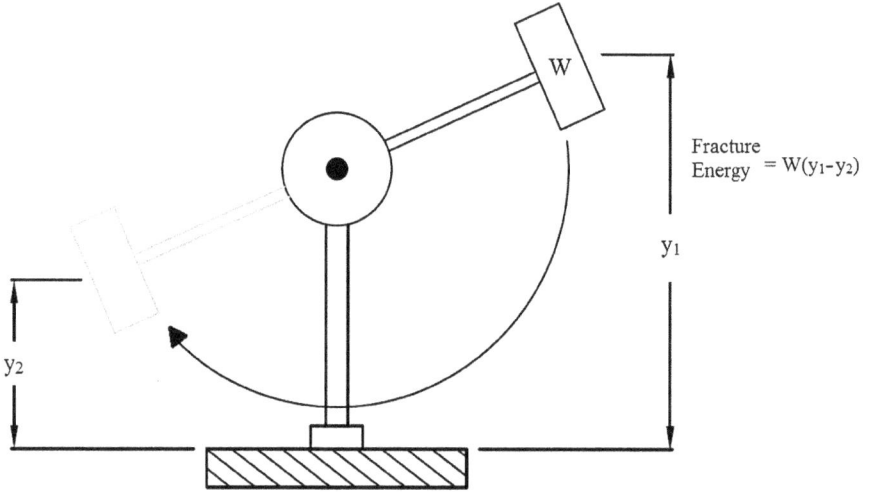

FIGURE 6.15   Sketch of impact testing machine.

The impact properties of most materials are affected by the temperature at which the test is performed. As such, test temperatures may characterize material behavior at either fixed values or over a range of temperatures. When testing over a range of temperatures, it is possible to characterize the transition region and the lower shelf and upper shelf behavior as shown in Figure 6.16. Upper shelf energies are associated with ductile fracture and lower shelf with brittle fracture.

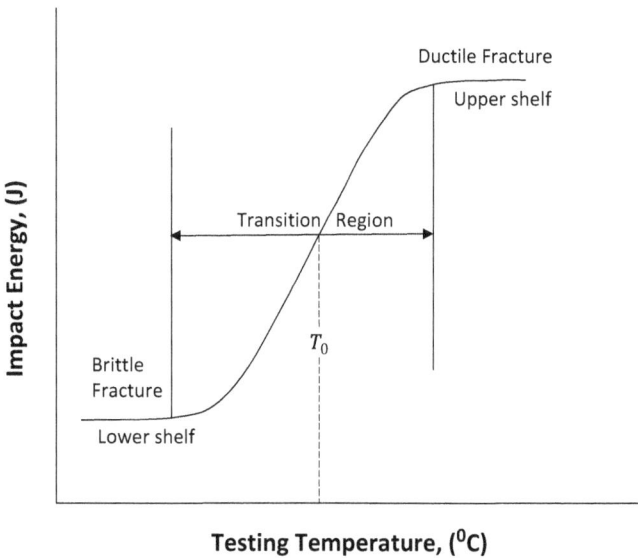

FIGURE 6.16   Charpy V-notch (CVN) impact energy for specimen at various test temperatures.

TABLE 6.1

Correlations of Charpy V-Notch Impact Energy to K_{IC}

Correlation	Temperature Range	Charpy V-Notch (CVN) Impact Energy	σ_{YS}	K_{IC}
$\dfrac{K_{IC}^2}{E} = 2\left(\text{CVN}\right)^{3/2}$ [7]	Transition	3–106 ft lb	39–246 ksi	87–246 ksi$\sqrt{\text{in}}$
$K_{IC} = 19(\text{CVN})^{1/2}$ [8]	Transition	—	303–820 MPa	—
$\left(\dfrac{K_{IC}}{\sigma_{YS}}\right)^2 = \dfrac{5}{\sigma_{YS}}\left(\text{CVN} - \dfrac{\sigma_{YS}}{20}\right)$ [7]	Upper shelf	16–89 ft lb	110–246 ksi	87–246 ksi$\sqrt{\text{in}}$
$\left(\dfrac{K_{IC}}{\sigma_{YS}}\right)^2 = 0.52\left(\dfrac{\text{CVN}}{\sigma_{YS}} - 0.02\right)$ [9]	Upper shelf	—	—	80–220 MPa$\sqrt{\text{m}}$

Several attempts have been made to correlate the Charpy energy to fracture toughness [6]. Although some of the empirical relationships appear to work fairly well, they are not always reliable and thus cannot be substituted for fracture mechanics based testing. Some of these empirical correlations for K_{IC} are given in Table 6.1 for certain steels. It is important to consider not only the type of material when selecting the appropriate relationship for K_{IC}, but also the temperature range, impact energy, and yield strength where applicable, for the material.

Example 6.4

A standard Charpy V-notch (CVN) impact test performed at room temperature on a specimen produced a measured impact energy of 35 J. If the specimen has a plane strain fracture toughness of 120 MPa$\sqrt{\text{m}}$, estimate K_{IC} using the impact energy and the appropriate relationship, given the following:
$\sigma_{YS} = 945$ MPa; $\sigma_{TS} = 1034$ MPa; $E = 207{,}000$ MPa

From Table 6.1, based on the impact energy, yield strength, and upper shelf energy relationship based on temperature, we have:

$$\left(\frac{K_{IC}}{\sigma_{YS}}\right)^2 = 0.52\left(\frac{\text{CVN}}{\sigma_{YS}} - 0.02\right)$$

$$K_{IC} = \sigma_{YS} \cdot \sqrt{0.52\left(\frac{\text{CVN}}{\sigma_{YS}} - 0.02\right)}$$

$$K_{IC} = \left(945\,\text{MPa}\right)\sqrt{0.52\left(\frac{35\,\text{J}}{945\,\text{MPa}} - 0.02\right)} = 89\,\text{MPa}\sqrt{\text{m}}$$

This is 25.8% lower than the actual K_{IC}.

Concept Challenge 6.4

If instead the transition temperature relationship was selected, would this provide an acceptable result?

6.3   DUCTILE TO BRITTLE TRANSITION TEMPERATURE TEST

6.3.1   STANDARD METHOD

The ASTM E1921-20 [10] outlines standard test procedures for determining a reference temperature T_0, which can be used to characterize the fracture toughness of ferritic steels that experience cleavage cracking. The test method is valid for steels with yield strengths ranging from 275 to 825 MPa. The fatigue pre-cracked specimens for the test can be single edged notched bend bars and standard or disk shaped compact tension specimens. In the ductile to brittle transition range, toughness data tends to be highly scattered making it difficult to determine a single value's toughness at a particular temperature. As a result, statistical methods are employed by E1921-20 to generate a transition toughness curve based on the master curve model which can be used for determining the transition temperature T_0.

The three-parameter Weibull distribution is used for defining a fracture's toughness at a fixed temperature in the transition region and is given by

$$P_f = 1 - \exp\left\{ -\left[\frac{K_{JC} - K_{min}}{K_0 - K_{min}} \right]^b \right\} \tag{6.24}$$

where

P_f = cumulative probability
K_{JC} = fracture toughness based on the critical J converted to the equivalent K, MPa$\sqrt{m}$
K_{min} = threshold toughness
K_0 = Weibull mean toughness taken at the 63.2 percentile toughness for a 25.4 mm (1 in) thick specimen
b = Weibull slope, characterizing the typical scatter of K_{JC} data

ASTM chooses to fix K_{min} at 20 MPa$\sqrt{m}$ with a Weibull slope of 4, when defining the model. This allows the unknown parameter K_0 to be fitted using the statistical distribution through a relatively small sample size.

6.3.2   INTERPRETATION OF RESULTS

Cleavage fracture is assessed by the weakest-link theory which provides a relationship between specimen size and K_{JC}. Toughness values for various thicknesses can thus be converted to equivalent values for a standard 1T specimen (25 mm thick) using

$$K_{JC(1T)} = K_{min} + \left(K_{JC(B)} - K_{min}\right)\left(\frac{B}{25}\right)^{1/4} \tag{6.25}$$

where B is specimen thickness in millimeters

Using these converted toughness values, K_0 can be computed using

$$K_0 = \left[\sum_{i=1}^{N} \frac{\left(K_{JC(i)} - K_{min}\right)^4}{N}\right]^{1/4} + K_{min} \tag{6.26}$$

where N is the number of valid specimens tested. For a fixed temperature, the median toughness $K_{JC(med)}$ for a 1T specimen is given by

$$K_{JC(med)} = K_{min} + 0.91\left(K_0 - K_{min}\right) \tag{6.27}$$

The master curve model relates the temperature dependence of $K_{JC(med)}$ in the ductile to the brittle transition region by

$$K_{JC(med)} = 30 + 70\exp\left[0.019(T - T_0)\right] \tag{6.28}$$

where T_0 is the reference transition temperature in degrees Celsius. Figure 6.17 illustrates a fracture toughness master curve for an A533B steel specimen.

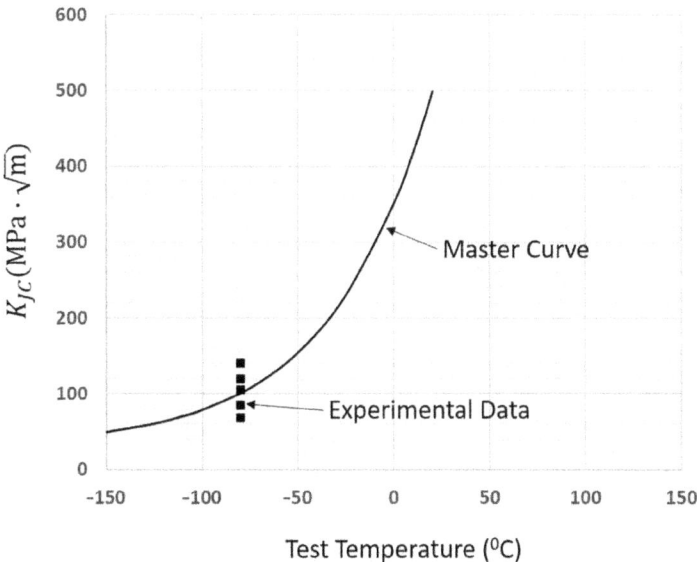

FIGURE 6.17   Fracture toughness master curve for 1T specimen.

Using the data to compute K_0 and $K_{JC(med)}$, the transition temperature is found using

$$T_0 = T - \left(\frac{1}{0.019}\right)\ln\left[\frac{K_{JC(med)} - 30}{70}\right] \tag{6.29}$$

ASTM recommends performing at least six fracture toughness tests.

Example 6.5

Fracture toughness testing is conducted on six compact specimens (1T) of A 533 steel at a test temperature of −75 °C. Determine the ductile to brittle transition temperature, if using the following test data:

Specimen #	1	2	3	4	5	6
$K_{JC(1T)}\left(MPa\sqrt{m}\right)$	80.1	90.0	103.9	115.1	123.6	140.9

We can use Eqn (6.26) with $K_{min} = 20\,MPa\sqrt{m}$ to find K_0 as follows:

$$K_0 = \left[\sum_{i=1}^{6}\frac{\left(K_{JC(i)} - K_{min}\right)^4}{N}\right]^{1/4} + K_{min}$$

$$K_0 = \left[\frac{(80.1-20)^4}{6} + \frac{(90.0-20)^4}{6} + \frac{(103.9-20)^4}{6} + \frac{(115.1-20)^4}{6}\right.$$

$$\left. + \frac{(123.6-20)^4}{6} + \frac{(140.9-20)^4}{6}\right]^{1/4} MPa\sqrt{m} + 20\,MPa\sqrt{m}$$

$$= 115.4\,MPa\sqrt{m}$$

Since the test temperature is fixed, we can use Eqn (6.27) to find $K_{JC(med)}$:

$$K_{JC(med)} = K_{min} + 0.91\left(K_0 - K_{min}\right)$$
$$= \left(20 + 0.91(115.4 - 20)\right)MPa\sqrt{m} = 106.8\,MPa\sqrt{m}$$

We can now use Eqn (6.29) to solve for T_0 as follows:

$$T_0 = T - \left(\frac{1}{0.019}\right)\ln\left[\frac{K_{JC(med)} - 30}{70}\right] = -75 - \left(\frac{1}{0.019}\right)\ln\left[\frac{106.8 - 30}{70}\right]$$
$$= -79.9°C$$

6.4   K-R CURVE

6.4.1   STANDARD METHODS

In the previously discussed section for K testing based on the ASTM E 399 standard, a single point value on the R curve was defined for K_Q. It is assumed that an unstable fracture occurs when the applied stress intensity factor has reached the material's resistance to fracture. However, in thin sections, a period of stable crack growth often occurs prior to unstable failure. The fracture toughness function for K_Q exhibits a size dependence since it contains the ligament length of the specimen as indicated in Eqn (6.2). The ASTM E 561-20 [11] standard provides an alternative to the single tough- ness measurement technique that involves defining the entire R curve for materials that exhibit ductile crack extension. This K-R curve method requires no minimum thickness, thus facilitating the testing of thin sheets. It does, however, require the plastic zone be small in comparison to the in-plane specimen dimensions.

Figure 6.18 shows a typical $K-R$ curve for a primarily linear elastic material. It provides a record of the toughness development as the crack is driven stably under an increasing applied stress intensity factor K.

Initially, there is very little crack growth with increasing K. Once the crack begins to grow, K increases with crack growth until the R curve flattens out. A critical stress intensity factor value, K_C, can be defined at the instability point where the driving force is tangent to the R curve. However, this value cannot be treated as a material property due to its dependence on the specimen size and geometry.

E561-20 describes the testing of middle-cracked tension M(T) or compact tension C(T) specimens for $K-R$ curve development. When testing thin sheets, anti-buckling plates are normally installed on either side of the specimen to limit out-of-plane buckling.

6.4.2   INTERPRETATION OF RESULTS

The $K-R$ curve is a plot of crack extension resistance K as a function of effective crack extension Δa_e. The standard describes several procedures for determining data pairs of K and Δa_e, including direct measurement of the physical crack size and

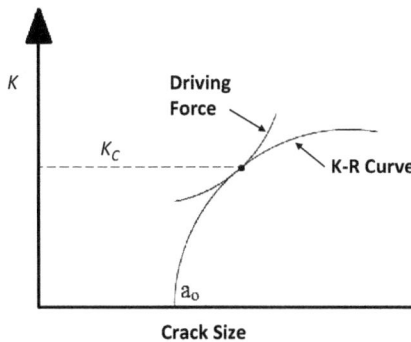

FIGURE 6.18   Sketch showing a $K-R$ curve.

specimen compliance techniques. When a plastic zone develops ahead of the growing crack the stress intensity should be corrected for plasticity effects by defining an effective crack length. ASTM suggests either the secant method or Irwin plastic zone correction. The Irwin approach defines the plastic zone size by r_y as discussed in Chapter 2. The effective crack size, a_{eff}, can then be defined as

$$a_{eff} = a_p + r_y \tag{6.30}$$

where a_p is the physical crack length measure visually. If the plastic zone is negligible, only the physical crack length measurement is needed. The instantaneous stress intensity factor can be determined using the appropriate stress intensity equation for the selected test specimen type. In the case of the middle tension (M(T)) specimen,

$$K = \frac{P}{BW} \cdot \sqrt{\pi a_{eff} \cdot \sec\left(\frac{\pi a_{eff}}{W}\right)} \tag{6.31}$$

where
   P is the applied force
   B is the specimen thickness
   W is the specimen width measured from the load line
   a_{eff} is the effective crack size which can be either a_e or physical crack size a_p as determined by the plastic zone assumption.

As variability in $K-R$ curves can be expected due to material type and mechanical properties, ASTM recommends duplicate tests be performed on multiple lots of materials, when developing design data.

Example 6.6

A fracture toughness test is conducted on an M(T) specimen using the ASTM E 561 procedure for $K-R$ testing. The effective crack length at a particular unloading compliance point was found to be 131.5 mm for an applied load of 479.4 kN. If $B = 6.8$ mm and $W = 760$ mm, estimate the corresponding K coordinate on the K_R curve.
   $B = 6.8$ mm and $W = 760$ mm, so

$$K = \frac{479.4 \times 10^{-3}\,\text{MN}}{(6.8 \times 10^{-3}\,\text{m})(0.76\,\text{m})} \cdot \sqrt{\pi(0.1315\,\text{m}) \cdot \sec\left(\frac{\pi(0.1315\,\text{m})}{0.76\,\text{m}}\right)}$$

$$K = 60\,\text{MPa}\sqrt{\text{m}}$$

PROBLEMS

6.1. K_{IC} testing performed according to ASTM E 399 standard using 1T compact specimens with $\sigma_{YS} = 600$ MPa results in a $K_Q = 60\,\text{MPa}\sqrt{\text{m}}$. If all validity requirements are met for the test, determine the required load capacity for the test machine, given $f(a/W) = 8.22$.

6.2. Load-displacement data obtained from a fracture toughness test on an aluminum alloy ($\sigma_{YS} = 68$ ksi) is given in the table below. A standard 1T compact tension specimen was used in accordance with the ASTM E 399 testing procedure.

Load (lbs)	Displacement (inches)
0	0
500	0.025
1000	0.05
1500	0.075
2000	0.1
3000	0.15
3350	0.18
3500	0.23
3400	0.27

Using the data, plot a load-displacement curve and use it determine P_Q by applying the 95% secant method. If possible, determine the fracture toughness for the alloy given:

$$B = 1.0 \text{ in}; W = 2.0 \text{ in}; a = 1.067 \text{ in}.$$

6.3. A 2024-T4 aluminum alloy is to be used in an ASTM E 399 fracture toughness test. Estimate the specimen dimensions required for a valid K_{IC} test given the following:

$$\sigma_{YS} = 324 \text{ MPa}; K_{IC} = 37 \text{ MPa}\sqrt{m}$$

6.4. Compliance testing is performed using an SE(B) specimen with $\sigma_{YS} = 345$ MPa and $\sigma_{TS} = 483$ MPa. $J_Q = 498$ kJ/m^2 is obtained from constructing the J-R curve according to the ASTM E 1820 procedure. Determine J_{IC}, if possible, given the following:

$$B = 25.4 \text{ mm}; W = 50.8 \text{ mm}; b_0 = 11.9 \text{ mm}; E = 210,000; \text{ MPa}; v = 0.3$$

6.5. A material having $K_{IC} = 110 \text{ MPa}\sqrt{m}$ and $\sigma_{YS} = 350$ MPa is to be used to determine J_{IC} using the ASTM E 1820 procedure. Determine the required thickness for the specimen, if $E = 207,000$ MPa, $\sigma_{TS} = 450$ Mpa, and $v = 0.3$.

6.6. A CTOD test is performed according to the ASTM E 1820 standard, on a structural steel specimen having $\sigma_{YS} = 410$ MPa, $\sigma_{TS} = 550$ Mpa, and dimensions of B = 25.4 mm and W = 50.8 mm, with a pre-crack to a depth of

27 mm. Stable crack growth is not present and J_{pl} is computed as 0.189 MJ/m^2. Determine δ, given the following:

$$K_{IC} = 101\,MPa\sqrt{m}; E = 210,000MPa; v = 0.3$$

6.7. An impact-testing machine with an 18 kg hammer and arm length of 125 cm (measured from the fulcrum point of impact) is initially set in the 90° position. Determine (a) the potential energy stored in the mass in the initial position; (b) the potential energy at this point if the hammer swings 45° (Figure 6.19) after impact with the specimen as shown; and (c) the fracture energy for the Charpy V-notch specimen.

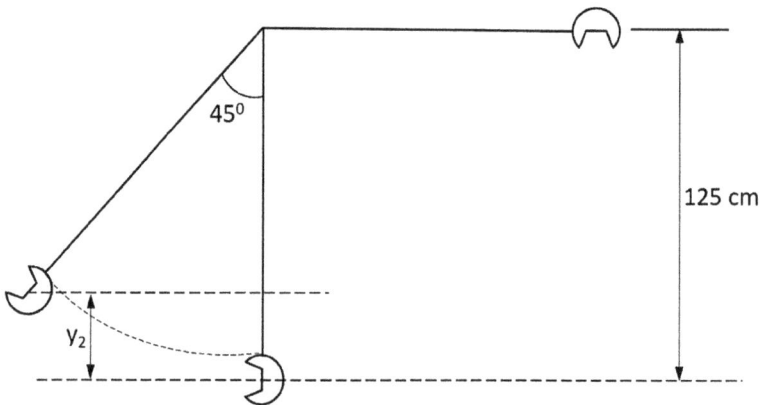

FIGURE 6.19   Impact test showing hammer position at critical locations.

6.8. A Charpy impact test performed on a low carbon steel alloy produced the following results:

Temperature (°C)	Impact Energy (J)
50	75
40	75
30	69
20	58
10	37
0	24
−10	15
−20	10
−30	4.5
−40	1.5

(a) Plot the impact energy versus temperature and using it to determine, (b) a ductile-to-brittle transition temperature using the average of the maximum and minimum impact energies, and (c) a ductile-to-brittle transition temperature for which the impact energy is 22 J.

6.9. A fracture toughness test is conducted at $-40\ ^\circ$C using eight 0.5T compact specimens. Determine the converted fracture toughness value for an equivalent 1T specimen for specimen #3 with $K_{JC(0.5T)} = 116.8\ \mathrm{MPa}\sqrt{\mathrm{m}}$.

6.10. The table below gives fracture toughness test data conducted on 0.5T compact specimens at $-55\ ^\circ$C. Complete the table for an equivalent 1T specimen and using the converted K_{JC} data determine the ductile to brittle transition temperature for the material if the ASTM E 1921-20 standard were employed for testing.

Specimen #	1	2	3	4	5	6	7	8
$K_{JC(0.5T)}\left(\mathrm{MPa}\sqrt{\mathrm{m}}\right)$	86.5	81.9	114.3	122.9	106.7	96.8	111.3	98.9
$K_{JC(1T)}\left(\mathrm{MPa}\sqrt{\mathrm{m}}\right)$								

6.11. An M(T) specimen with $B = 3$ mm, $W = 406$ mm, and $\sigma_{YS} = 533$ MPa used in K_R testing produces the following data:

$K_R\left(\mathrm{MPa}\sqrt{\mathrm{m}}\right)$	Crack Extension, mm
0	25.3
0.1	51.4
0.11	56.1
0.13	62.7
0.21	76.1
1.1	101.6
2.3	114.7
5.7	130.7
9.6	142.1
14.8	152.4
23	160
30	162.8
38	164.7

(a) Plot the K_R for the given data.

(b) Determine the load required to produce a crack extension of 6.5 mm.

6.12. Estimate the plastic zone size for a M(T) specimen with an observed physical crack length of 21.7 mm during K_R testing. The specimen has $B = 12.5$ mm, $W = 254$ mm, and $\sigma_{YS} = 314$ MPa.

REFERENCES

[1] P. F. Filgueiras, J. T. Oliveira de Menezes and J. E. P. Ipiña, "Fracture toughness of high strength seamless pipe steel from SE(T) and SE(B) specimens evaluated by different standards," *Fatigue & Fracture of Engineering Materials & Structures*, pp. 572–582, 2018.

[2] E1823-20b, "Standard Terminology Relating to Fatigue and Fracture Testing," *American Society for Testing and Materials*, 2020.

[3] E399-20a, "Standard Test Method for Linear-Elastic Plane-Strain Fracture Toughness of Metallic Materials," *American Society for Testing and Materials*, 2020.

[4] E1820-20b, "Standard Test Method for Measurement of Fracture Toughness," *American Society for Testing and Materials*, 2020.

[5] E23-18, "Standard Test Methods for Notched Bar Impact Testing of Metallic Materials," *American Society for Testing and Materials*, 2018.

[6] G. T. Mendez, S. I. Colindres, J. C. Velazquez, D. A. Herrera, E. T. Santillan and A. Q. Bracarense, "Fracture Toughness and Charpy CVN Data for A36 Steel with Wet Welding," *Soldagem & Inspeção*, vol. 22, no. 3, pp. 258–268, 2016.

[7] J. M. Barsom and S. T. Rolfe, "Correlations between KIC and Charpy V-notch test results in the transition-temperature range," *ASTM STP 466*, pp. 281–302, 1970.

[8] B. Marandet and G. Sanz, "Evaluation of the toughness of the medium-strength by using elastic fracture mechanics and correlations between KIC and Charpy V-notch," *Flaw Growth and Fracture*, vol. ASTM STP 631, pp. 72–95, 1976.

[9] R. Robert and C. Newton, "Interpretive Report on Small Scale Test Correlations with KIC Data," *WRC*, vol. 265, pp. 1–16.

[10] E1921-20, "Standard Test Method for Determination of Reference Temperature, To, for Ferritic Steels in the Transition Range," *American Society for Testing and Materials*, 2020.

[11] E561-20, "Standard Test Method for KR Curve Determination," *American Society for Testing and Materials*, 2020.

7 Software Applications for Linear Elastic Fracture Mechanics

OBJECTIVES

After studying the first part of this chapter (Section 7.1), the student should be able to:

1. Describe the input options needed for linear elastic fracture mechanics (LEFM) static analysis.
2. Calculate the factors of safety for elementary LEFM problems by hand.
3. Use a basic fracture mechanics calculator to solve static LEFM problems.
4. Understand the assumptions associated with software inputs.

7.1 CRACK GROWTH SOFTWARE (LEFM STATIC APPLICATIONS)

There are several inputs associated with solving crack-growth problems when using software; these include but are not limited to: (i) material properties ($K_{Ic}, \sigma_{YS}, \sigma_{ULT}, E$), (ii) the model used (e.g. a through crack in the center of plate), (iii) the applied load, and (iv) crack geometrical values. Figure 7.1 provides an example of a fracture mechanics calculator available at mechanicalc.com [1] and provides an image of the geometric input.

Example 7.1

A center cracked panel with a 1 in. wide crack is subjected to a uniform tensile stress of 40 ksi and uniform bending stress of 10 ksi. The plain strain fracture toughness of the material is $50 \, \text{ksi}\sqrt{\text{in}}$; its yield stress is 80 ksi, its tensile strength is 120 ksi, and its Young's Modulus is 3×10^4 ksi; the plate width and thickness are 14 in and 2 in. respectively. Calculate the LEFM Factor of Safety.

We will use the fracture mechanics calculator at Mechanicalc.com; however, other fracture mechanics (FM) software will be similar.

INPUTS

The fracture mechanics input widow at mechanicalc.com will display options shown in Figures 7.1 and 7.2. For the crack type, we select "center through crack." The geometric and applied load inputs are listed below and entered as shown in Figures 7.1 and 7.2.

GEOMETRIC PROPERTIES

Crack length, $a = 0.5$ in
  Plate half-width, $b = 14$ in (typically W in this text)

DOI: 10.1201/9781003052050-7

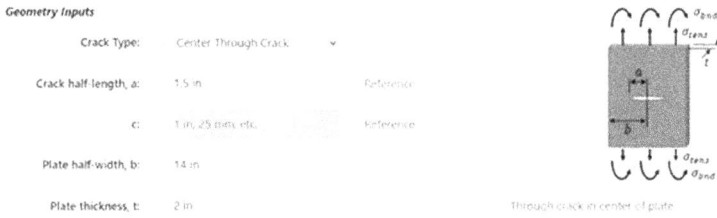

FIGURE 7.1 Fracture mechanics calculator input (Image Source: MechaniCalc.com).

Thickness, $t = 2$ in (typically B in this text)
Tensile stress, $\sigma_{tens} = 40$ ksi
Bending stress, $\sigma_{bnd} = 10$ ksi

Since an actual material is not provided, a material will have to be created. The minimum properties needed for an LEFM static analysis are the fracture toughness K_{Ic}, yield strength σ_{ys}, and Young's Modulus E. On inspection of Figure 7.2, we see an option in the Material Input section to "Create New." Selecting this option yields the windows in Figures 7.3 and 7.4. Note the asterisk by the necessary inputs. Specimen orientation refers to L-T, T-L, etc. as discussed in Chapter 6. A_k, B_k are the material constants in Eqn. (2.16); these are only necessary if plane stress K_c is needed and only the plane strain K_{Ic} is known. Though we are conducting a static analysis, fatigue crack growth (da/dN) information is needed to fully define the material. This entry is shown in Figure 7.4. The minimum fatigue property information needed are C_0, the intercept constant, and n, the slope on a log-log scale for the Paris–Ergodan model. We therefore choose a material name, and enter the Young's modulus, fracture toughness and yield strength to create our new material.

FACTOR OF SAFETY

With the material now defined, selecting "Calculate Results" (Figure 7.2) now provides the LEFM FS. In this problem

$$FS = \frac{K_{Ic}}{K_I} = 1.06$$

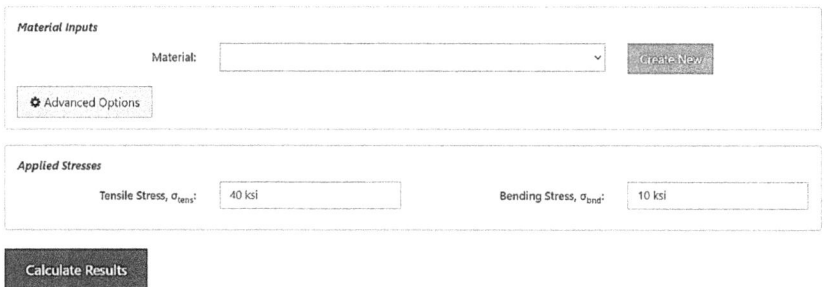

FIGURE 7.2 Material creation and stress input (Image Source: MechaniCalc.com).

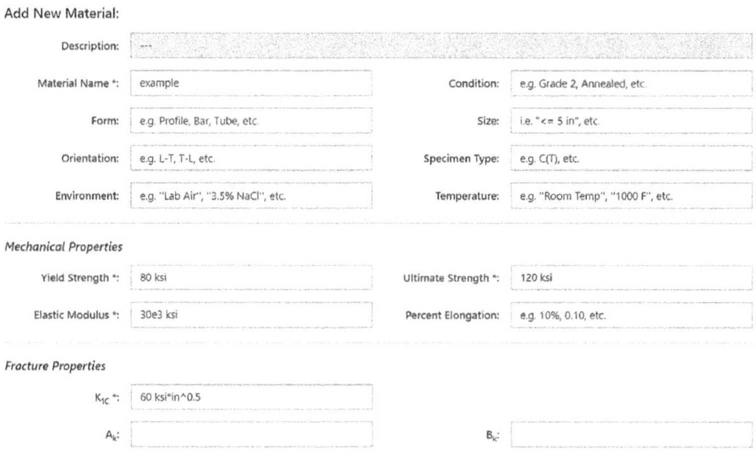

Add New Material:

Description: ...

Material Name *:	example	Condition:	e.g. Grade 2, Annealed, etc.
Form:	e.g. Profile, Bar, Tube, etc.	Size:	i.e. "<= 5 in", etc.
Orientation:	e.g. L-T, T-L, etc.	Specimen Type:	e.g. C(T), etc.
Environment:	e.g. "Lab Air", "3.5% NaCl", etc.	Temperature:	e.g. "Room Temp", "1000 F", etc.

Mechanical Properties

Yield Strength *:	80 ksi	Ultimate Strength *:	120 ksi
Elastic Modulus *:	30e3 ksi	Percent Elongation:	e.g. 10%, 0.10, etc.

Fracture Properties

K_{1C} *:	60 ksi*in^0.5		
A_k:		B_k:	

FIGURE 7.3 Material inputs with applied static loads (Image Source: MechaniCalc.com).

Fatigue Crack Growth Properties

C_0 *:	1E-17	(in/cycle) / (psi*in$^{0.5}$)^n	n *:	1
p:			q:	
$\Delta K_{th,1}$:	e.g. 1 ksi*in^0.5, 1 MPa*m^0.5, etc.		

Add Material

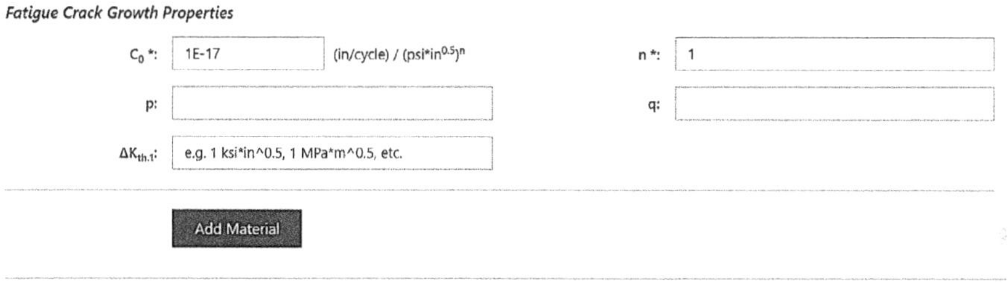

FIGURE 7.4 Material fatigue crack growth properties (Image Source: MechaniCalc.com).

CALCULATION

The geometric factors Y_T, Y_B for tensile and bending applied loads in a center-cracked panel can be found in any FM database; they are 1 and 0.5, respectively.

The stress intensity factor is calculated using superposition of the tensile and bending cases

$$K_I = \left(\sigma_T Y_T + \sigma_B Y_B\right)\sqrt{\pi a} \tag{7.1}$$

$$K_I = \left(40(1)+10(0.5)\right)\sqrt{(\pi \times 1.5)} = 56.4 \text{ ksi}\sqrt{in}$$

Is LEFM applicable?
From Eqn. (2.29),

$$2.5\left(\frac{K_I}{\sigma_{ys}}\right)^2 = 2.5\left(\frac{56.4}{80}\right)^2 = 1.24 \text{ in}$$

The distance from the crack tip to the boundary, $W - a = 14$ in $- 0.5$ in $= 13.5$ in > 1.24 in, similarly $a = 1.5$ in > 1.24 in, therefore, LEFM is applicable.

Plane strain vs. plane stress?

$B = 2$ in > 1.24 in, therefore a plane strain condition should be assumed.

The factor of safety can be calculated by:

$$FS = \frac{K_{Ic}}{K_I} = \frac{60}{56.4} = 1.06$$

Example 7.2

A 5 cm edge crack was discovered in the HY-80 steel T-beam shown in Figure 7.5, that is one of several T-beams structurally supporting a parking deck. The T-beam is subjected to both uniform transverse and axial loads. The loads result in a tensile stress of 35 MPa and a bending stress of 18 MPa. The web depth is 0.6 m and the web thickness 0.2 m. The rolling direction is along the longitudinal axis. What orientation is the fracture in ASTM notation? Show that LEFM is applicable to this problem. If yes, what is the LEFM Factor of Safety? What is the Factor of Safety for the critical stress and the critical crack length? How long would the crack have to be for this beam to fail via unstable crack growth?

SOLUTION

The loading on the crack is in the longitudinal orientation and crack propagation is in the transverse direction, so this is an L-T configuration.

After selecting the FM calculator at Mechanicalc.com [1], the following steps can be performed:

Step 1 Crack type, geometry, and loading

For the crack type, we select "single edge through crack." The geometric and applied load inputs are shown below and would be entered as before (see Figure 7.1), but this time in SI units.

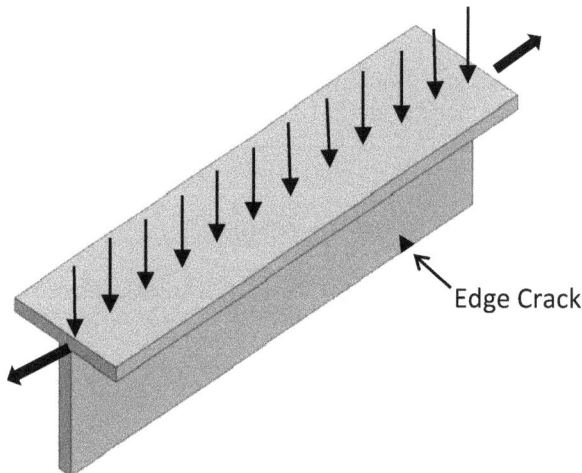

FIGURE 7.5 T-beam under uniform transverse load and axial load.

Crack length, a = 50 mm
Plate width, b = 600 mm (W in this text), the web depth for this problem
Plate Thickness, t = 200 mm (B in this text), the web thickness for this problem
Tensile stress, σ_{tens} = 35e3 kPa
Bending stress, σ_{bnd} = 18e3 kPa

Step 2 Material properties

"HY-80 steel L-T, T-L, Room air, Temp" steel is in the database for this FM calculator. If not available as an option in the pull-down menu, it can be added by selecting the "Create New" option (Figure 7.2) then choosing the "Materials Database"
tab.
    After selecting "HY-80 steel L-T, T-L, Room air, Temp" as the material for this
problem, selecting "Calculate Results" now provides the LEFM FS. The residual
strength curve is also plotted and Factors of Safety for the critical crack length and
critical stress provided.

Step 3 Software calculation

Since this material's properties are known, the software can plot its fracture toughness vs. thickness curve using Eqn. (2.16).
    From Figure 7.6 it is seen that the thickness falls in the plane stress region which
infers we could possibly use the plane stress fracture toughness in our analysis.
However, since the plastic zone size is the primary driver regarding choice of
plane stress vs. plane strain, we still need to check the limitation placed on the
thickness (based on the plastic zone size).
    *Geometry correction factor as a function of crack length for finite width plates
(Y vs. a/W).* Recall Y also varies with a/W; this software plots the graph using
standard accepted relationships shown in Table 2.2. The plot of Y vs. a is shown
in Figure 7.7; note that using a/W instead would have served to keep the x-axis
values between 0 and 1. Y_t and Y_B are found to be 1.189 and 1.049 respectively.
    The stress intensity factor is calculated by Eqn. (7.1), i.e.

$$K_I = \left(35(1.189)+18(1.045)\right)\sqrt{\left(\pi \times 5 \times 10^{-2}\right)} = 23.95\,MPa\sqrt{m}$$

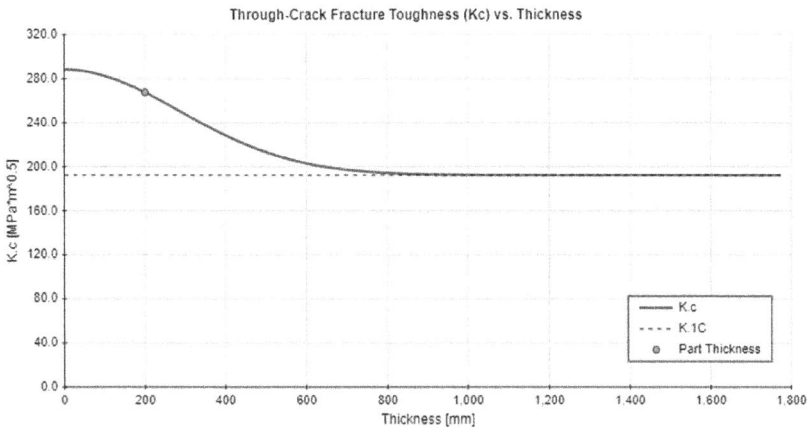

FIGURE 7.6 *Kc* vs. thickness for HY-80 steel in Example 7.2. (Image Source:
MechaniCalc.com).

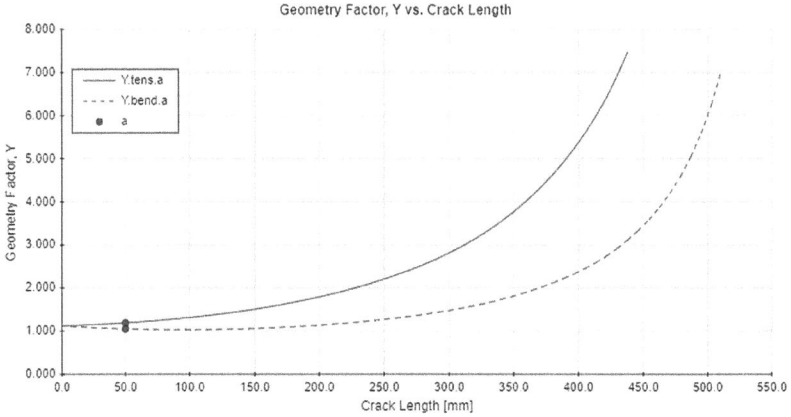

FIGURE 7.7 Y vs. a for edge crack under bending and tensile loads for Example 7.2 (Image Source: MechaniCalc.com).

We can check if LEFM is indeed applicable by examining whether

$$a, W - a > 2.5 \left(\frac{K_I}{\sigma_{ys}} \right)^2$$

$$2.5 \left(\frac{K_I}{\sigma_{ys}} \right)^2 = 2.5 \left(\frac{23.95}{551.6} \right)^2 = 4.7 \, \text{mm}$$

a, $W - a$ = 50 mm, 550 mm respectively, so LEM is applicable.
Do we use plain strain or plain stress fracture toughness?
We see the thickness B > 4.7 mm, so plane strain is recommended.
From the materials database, the plane strain fracture toughness for "HY-80 steel L-T, T-L, Room air, Temp" is 192.3 $MPa\sqrt{m}$, therefore

$$FS = \frac{K_{Ic}}{K_I} = \frac{192.3}{23.95} = 8.03$$

RESIDUAL STRENGTH CURVE

The residual strength curve (σ vs. a) is derived from Eqn. (7.2) and is plotted by the software as shown in Figure 7.8.
The combined stress, σ_C, is given by

$$\sigma_C = \frac{K_{Ic}}{Y(a/W)\sqrt{\pi a}} = \frac{192.3}{Y(a/W)\sqrt{\pi a}}$$

The point defined by the combined stress, σ_C, and crack length, a, can be plotted on the residual strength curve. A vertical line through this point intersects the residual strength curve to yield σ_{crit} = 409.1 MPa; the horizontal line through this point intersects the residual strength curve to yield a_{crit} = 364.2 mm.

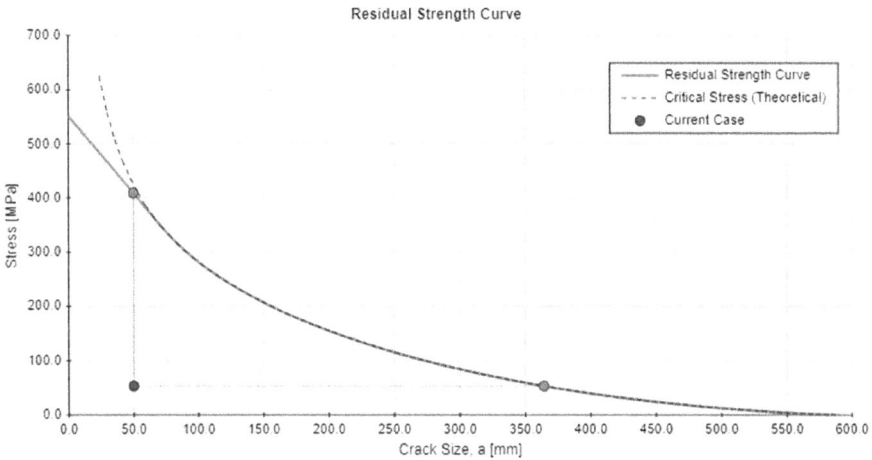

FIGURE 7.8 Residual strength curve (Example 7.2) (Image Source: MechaniCalc.com).

Therefore, the FS for residual strength is found by
Critical stress approach:

$$FS = \frac{\sigma_{crit}}{\sigma_c} = \frac{409.1}{53} = 7.72$$

Critical crack length approach:

$$FS = \frac{a_{crit}}{a} = \frac{364.2}{50} = 7.28$$

Example 7.3

A 7075-T6 aluminum plate of width 800 mm and thickness 100 mm is found to have a corner surface crack $a = 7$ mm, $c = 70$ mm, as shown in Figure 7.8. While in service, the plate is subjected to static tensile stress of 105 MPa. The crack is observed to grow in the a direction over time but remains stationary in the c direction. What is the critical crack length in the a direction for which failure will occur?

SOLUTION

Using the FM Calculator at Mechanicalc.com:

Step 1 Crack type, geometry, and loading

For the crack type, we select "corner surface crack." The geometric and applied load inputs are shown below and would be entered as before (see Figure 7.1), but this time in SI units.
Crack length, $a = 7$ mm
Plate width, $b = 800$ mm (W in this text)

Plate Thickness, t = 100 mm (B in this text)
Tensile stress, σ_{tens} = 80e3 kPa

Step 2 Material properties

Aluminum 7075 T6 is in the database for this FM calculator. If not available as an option in the pull-down menu, it can be added by selecting the "Create New" option (Figure 7.2) then choosing the "Materials Database" tab. Select "Aluminum 7075-T6."

Step 3 Calculate results

Select "Calculate Results" (see Figure 7.2).
The residual strength factor of safety shows that a_{crit} = 10.96 mm. If we were to vary the a value until the residual strength curve FS for the critical crack length equals one, we obtain a slightly higher a_{crit} = 11.55 mm. The lower a_{crit} should be used as a conservative estimate.

STATIC LOAD PROBLEMS

7.1. List and describe four material properties typically needed as input to solve basic LEFM static problems using software?

7.2. A finite-width plate (W = 14 cm) has a center-through-crack of length 8 cm ($2a$) when subjected to a tensile stress of 32 MPa. Show that LEFM applies to this problem. Determine the LEFM factor of safety by hand. See Appendix B for the related stress intensity formula. K_{IC} = 27 MPa$\sqrt{m}$, σ_{YS} = 35 MPa.

7.3. A long circular rod made from material with properties K_{Ic} = 25 MPa $\sqrt{m}$, σ_{YS} = 52 MPa, is subjected to a 100 MPa tensile stress. Use a standard mathematical program or a spreadsheet to plot the residual strength in the form of $\sigma\sqrt{W}$ vs. a/W. The relevant stress intensity relationship can be found in Appendix B. If W = 5 cm, and a circumferential crack of length a = 1 cm exists, determine the residual strength Factors of Safety for the critical crack length and critical stress from the graph.

7.4. A center cracked panel with a 0.3 in. wide crack is subjected to loads that result in a tensile stress of 36 ksi and a bending stress of 8 ksi. The plain strain fracture toughness of the material is 35 ksi$\sqrt{in}$. The panel's yield stress is 90 ksi, its tensile strength 130 ksi, and its Young's Modulus 3×10^6ksi; the plate width and thickness are 20 in. and 4 in. respectively. Calculate the LEFM Factor of Safety by hand.

Use a Fracture Mechanics Calculator or develop your own code to solve the following problems.

7.5. A 2.6 cm edge crack was discovered in an AISI 316 stainless steel, beam rectangular cross-section. The crack is perpendicular to the longitudinal axis and located on the side where the bending moment results in a tensile stress. The beam is subjected to a three-point bending load and axial loads. These loads result in a tensile stress of 56 MPa and a bending stress of 25 MPa at the crack location. The beam height is 1.2 m and the thickness is 0.3 m. The rolling direction is along the longitudinal axis. What orientation is the fracture in ASTM notation? Show whether LEFM is applicable to this problem.

If yes, what is the LEFM FS? What is the FS for the critical stress and the critical crack length? How long would the crack have to be for this beam to fail via unstable crack growth?

7.6. A 15-5PH stainless steel plate of width 600 mm and thickness 75 mm is found to have a corner surface crack $a = 3$ mm, $c = 55$ mm. While in service, the plate is subjected to static tensile stress of 125 MPa. The crack is observed to grow in the a direction over time but remains stationary in the c direction. What is the critical crack length in the a direction for which failure will occur?

OBJECTIVES

After studying the second part of this chapter (Sections 7.2–7.5), the student should be able to:

1. Solve elementary fatigue crack growth (FCG) problems using ubiquitous mathematical programming languages.
2. Describe the input options typically needed for an elementary LEFM FCG analysis when using FCG specific software.
3. Calculate the LEFM Factors of Safety for elementary LEFM FCG problems by hand.
4. Know and understand the options available within the software to conduct FCG analysis for problems that may be encountered in industry.

7.2 SOFTWARE USE IN LEFM FCG ANALYSIS

The fatigue life is usually defined as the number cycles needed for a crack (in the part) to grow from some initial finite size to its critical length, while the part is subjected to fatigue loading. A general algorithm that calculates the number of cycles until failure is shown in Figure 7.9. It assumes constant amplitude loading, and that the Paris equation appropriately models the da/dN vs. ΔK behavior.

Basic FCG problems are usually algebraic, iterative, and incorporate single variable calculus. These problems therefore lend themselves well to mathematical coding languages which are standard throughout the engineering industry. For basic problems, the Paris equation is usually a sufficient model for the crack growth behavior. The Paris equation was described in Chapter 4 and is repeated here for convenience. It defines the straight-line region (region II) of the da/dN vs. ΔK curve.

$$\frac{da}{dN} = C(\Delta K)^n \tag{7.2}$$

The Paris equation does not account for the stress ratio R. As shown in Chapter 4, the crack growth rate depends on the stress ratio; therefore when the Paris equation is used, the value of C used will have to be specific to the particular stress ratio for the problem. A basic FCG problem is provided in Example 7.4, followed by the MATLAB and Mathematica solutions.

```
          ┌──────────────────────────────────┐
          │     Start with a₀, Δσ, ΔN        │
          └──────────────────────────────────┘
                           │
                           ▼
          ┌──────────────────────────────────┐
    ┌────▶│     K_max = Yσ_max√πa            │
    │      └──────────────────────────────────┘
    │                      │
    │                      ▼
    │              ┌────────────────┐
    │              │  Is K_max ≥ K_c?│
    │              └────────────────┘
    │               │              │
    │               ▼              ▼
    │   ┌──────────────────┐   ┌──────────┐
    │   │  ΔK = YΔσ√πa     │   │ Failure  │
    │   └──────────────────┘   └──────────┘
    │            │
    │            ▼
    │   ┌──────────────────┐
    │   │  da/dN = C(ΔK)ⁿ  │
    │   └──────────────────┘
    │            │
    │            ▼
    │   ┌──────────────────┐
    │   │  Δa = da/dN ΔN   │
    │   └──────────────────┘
    │            │
    │            ▼
    │   ┌──────────────────┐
    └───│  a_{i+1} = a_i + Δa│
        │  N_{i+1} = N_i + ΔN│
        └──────────────────┘
```

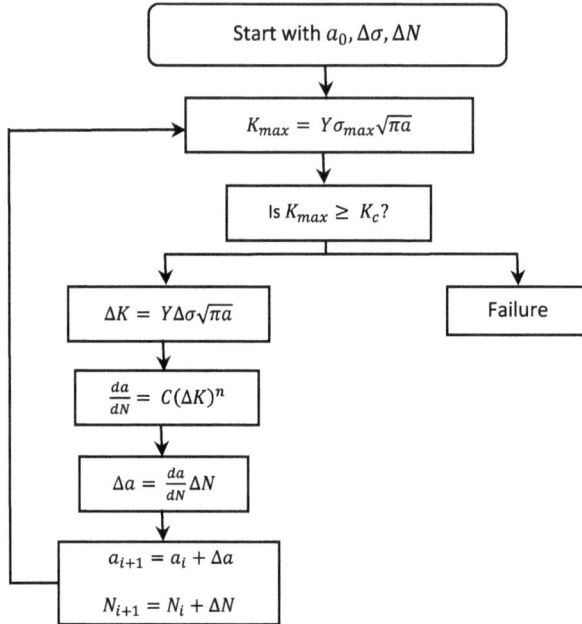

FIGURE 7.9   General algorithm for LEFM fatigue analysis (Image Source: MechaniCalc.com).

Example 7.4

A large edge cracked plate with fracture toughness 75 MPa containing an initial crack of length $a = 20$ mm is subjected to a constant amplitude cyclic tensile stress ranging between a minimum value of 50 MPa and a maximum of 185 MPa. Assuming the fatigue crack growth rate a is governed by the equation $da/dN = 0.51 \times 10^{-11}(\Delta K)^3$, calculate the crack growth rate when the crack length has the following values: 20 mm, 30 mm, 40 mm. Estimate the number of cycles to failure.

For the edge-cracked semi-infinite plate in tension, the stress intensity factor is given by $K = 1.12\sigma\sqrt{\pi a}$; note we have dropped the Mode I subscript for simplicity.

For a = 20mm,

$$\Delta K = 1.12\,\Delta\sigma\sqrt{\pi a}$$
$$= 1.12(185-50)\sqrt{\pi(0.02)}$$
$$= 37.9\ MPa\sqrt{m}$$

$$\frac{da}{dN} = 0.51\times10^{-11}(37.9)^3 = 2.78\times10^{-7}\ m/cycle$$

$\dfrac{da}{dN}$ for $a = 30$ mm is found in a similar manner, yielding $da/dN = 5.1 \times 10^{-7}$ m/cycle and for $a = 40$ mm, $\dfrac{da}{dN} = 7.85 \times 10^{-7}$ m/cycle

The critical crack size is found by

$$a_{crit} = \frac{1}{1.12\pi}\left(\frac{K_{Ic}}{\sigma_{max}}\right)^2 = \frac{1}{1.12\pi}\left(\frac{75}{185}\right)^2 = 46.7\,\text{mm}$$

$$N_f = \int dN = \int_{2\times10^{-2}}^{4.67\times10^{-2}} \frac{da}{0.51\times10^{-11}(\Delta K)^3}$$

$$= \int_{2\times10^{-2}}^{4.67\times10^{-2}} \frac{da}{0.51\times10^{-11}\left(1.12\times135\times\sqrt{\pi a}\right)^3}$$

$$= \int_{2\times10^{-2}}^{4.67\times10^{-2}} \frac{1.0187\times10^4}{a^{3/2}}da = 49797\ cycles$$

The process above could be programmed in mathematical software such as MATLAB, PYTHON, Mathematica, or a spreadsheet such as Excel.

MATLAB

The problem coded in MATLAB is shown below:

```
ainit=input('Enter initial crack length value in meters = ');
SigmaMax=input('Enter Max. Tensile Stress value (MPa) = ');
SigmaMin=input('Enter Min. Tensile Stress value (MPa)= ');
dadNcoeff=input('Enter da/dN coefficient = ');
KIC=input('Enter Fracture Toughness value (MPa_sqrt(m))= ');
DeltaSigma=(SigmaMax-SigmaMin);
syms a   % defines the crack length as a symbolic variable
DeltaK=(1.12*DeltaSigma*sqrt(pi*a));
dadN = (dadNcoeff)*(DeltaK)^3;
acrit=(1/(1.12*pi))*(KIC/SigmaMax)^2
f = 1/dadN; % define the function to be integrated
I = int(f,ainit,acrit); % integrates the function
Number_of_cycles = round(vpa(I)) % rounds to nearest
integer
```

The MATLAB output is shown in Figure 7.10

```
Command Window
    Enter initial crack length value in meters = 0.02
    Enter Maximum Tensile Stress value (MPa) = 185
    Enter Minimum Tensile Stress value (MPa)= 50
    Enter da/dN coefficient = 0.51*10^-11
    Enter Fracture Toughness value (MPa_sqrt(m))= 75

    acrit =

        0.0467


    Number_of_cycles =

    49797

fx >>
    <
```

FIGURE 7.10 MATLAB output for Example 7.4.

7.3 FCG SPECIFIC SOFTWARE

For real world cases, more specific FCG software may be needed for a variety of reasons. These reasons include but are not limited to: (i) there are different da/dN vs. ΔK relationships other than the Paris equation that may be more suitable, depending on the problem; (ii) the da/dN vs. ΔK data at the relevant R-ratio may not be available; (iii) the amplitude of the loading could be variable; (iv) the load spectra may have overloads and underloads; and (v) there may be environmental data available.

The typical inputs required when solving an FCG problem using a software program include but are not limited to: (i) the material, (ii) the model, and (iii) the load or stress spectrum. We will use the FM calculator at mechanical.com [1] to demonstrate the input process.

7.3.1 MATERIAL OPTION

When defining the material for a fatigue analysis, the same material properties required for a static LEFM analysis, i.e. $(K_{Ic}, \sigma_{YS}, \sigma_{ULT}, E)$, will be needed as well as the fatigue properties relevant to the da/dN vs. ΔK model choice. The material input for the FCG calculator at mechanicalc.com is the same as that required for the FM calculator shown in Figure 7.4. Recall that at least two FCG model parameters (C_0, n) were required to fully define the material, even for a static analysis only. For materials already defined in the system, the software will have the da/dN vs. ΔK curves at various R-ratios and/or the values of the fatigue parameters for selected da/dN vs. ΔK models.

The equation used by this FCG calculator to model the da/dN vs. ΔK curve is known as the NASGRO equation. The NASGRO equation [2, 3] is used in NASA's crack growth life prediction software, NASGRO, Version 3.0 (at the time of writing). It is the most general of the crack growth equations, accounting for the stress ratio R and crack closure effects. It also models all three regions of the crack growth rate curve. The NASGRO equation is described below; it will be an option for most, if not all FCG software.

7.3.2 NASGRO Equation

The NASGRO equation is shown in Eqn. (7.3). The equations that define and relate the empirically derived parameters in the NASGRO equation are shown in Eqn. (7.4–7.9). Table 7.1 lists the NASGRO parameters and their definitions.

$$\frac{da}{dN} = C\left[\left(\frac{1-f}{1-R}\right)\Delta K\right]^n \frac{\left(1-\dfrac{\Delta K_{th}}{\Delta k}\right)^p}{\left(1-\dfrac{K_{max}}{K_{crit}}\right)^q} \qquad (7.3)$$

$$f = \frac{K_{op}}{K_{max}} = \begin{cases} max\left(R,\, A_0 + A_1 R + A_2 R^2 + A_3 R^3\right) R \geq 0 \\ A_0 + A_1 R - 2 \leq R \leq 0 \\ A_0 - 2A_1 \ R < -2 \end{cases} \qquad (7.4)$$

$$A_0 = \left(0.825 - 0.34\alpha + 0.05\alpha^2\right)\left[\cos\frac{\pi\sigma_{max}}{2\sigma_0}\right]^{1/\alpha} \qquad (7.5)$$

$$A_1 = \left(0.415 - 0.071\alpha\right)\left(\frac{\sigma_{max}}{\sigma_0}\right) \qquad (7.6)$$

$$A_2 = 1 - A_0 - A_1 - A_3 \qquad (7.7)$$

$$A_3 = 2A_0 + A_1 - 1 \qquad (7.8)$$

TABLE 7.1
NASGRO Parameters

a	Crack Length
a_0	Intrinsic crack length (0.015 in or 0.0000381 m)
C_0	The intercept constant, C, for the case where the stress ratio $R = 0$
R	Stress ratio
f	Newman crack closure function (also called crack opening function)
α	Constraint factor, $\alpha = 1$ for plane stress, $\alpha = 3$ for plane strain
C_{th}	Threshold coefficient
ΔK_{th}	Threshold of stress intensity range for crack propagation
ΔK_0	Threshold stress intensity range at $R = 0$
n	Slope on a log-log scale
ΔK	Stress intensity range
K_c	Critical stress intensity or plane strain fracture toughness if thick section
σ_0	Flow stress, the average of the yield and ultimate stresses
C, p, q	Empirical constants

$$\Delta K_{th} = \Delta K_0 \left(\frac{a}{a+a_0}\right)^{1/2} \Big/ \left(\frac{1-f}{(1-A_0)(1-R)}\right)^{1+C_{th}R} \qquad (7.9)$$

A_0, A_1, A_2 A_3 C_{th}, ΔK_0 values are available in the NASGRO material database for each material. In applying the NASGRO equation, some software will request $\Delta K_{th.1}$; this is the threshold stress intensity at $R = 1$.

7.3.3 MODEL

The model refers to the type of crack or crack configuration (e.g. internal through crack, surface edge crack) for a given geometry (pipe, plate, hole, etc.). The same model options available for FM calculators are typically available for FCG calculators. Model options for the FCG calculator at mechanicalc.com include: Center Through Crack, Single Edge Through Crack, Elliptical Surface Crack, Corner Surface Crack, and Thumbnail Crack in Cylinder.

7.3.4 LOAD INPUT

The FCG calculator at mechanicalc.com will be used as an example here. For this FCG calculator, the stresses are entered directly, as shown in Figure 7.12. In this case, two stress ranges, i.e. $(\sigma_{max}, \sigma_{min})$, and the number of cycles, N, for each stress range have been entered.

Example 7.5

During inspection of a 7075-T6 aluminum panel, a center through crack of length 0.2 inches was found. The panel is one of several that constitute the skin of an aircraft fuselage. Upon return to service, the fuselage will be subjected to a series of constant amplitude cyclical hoop stresses before the next maintenance event at 20,000 cycles. The stress spectra generated is a repetition of the stress cycles below.

1.   15 *ksi* and −5 *ksi* respectively for 3,000 flight cycles
2.   3.5 *ksi* and 0 *ksi* for 7,000 flight cycles.

Determine whether the panel will fail before the next maintenance event. The panel width and thickness are 40 in. and 1 in. respectively. What if the load sequence were reversed?

We will use the FCG calculator at mechanicalc.com[1] to solve this problem.

After selecting the FCG calculator at mechanicalc.com, the following steps can be performed:

Step 1 Geometry and load inputs

For the model, we select "center through crack." The geometric and applied stress inputs are shown below and entered in the same manner as they were entered for the static case (see Examples 7.1 and 7.2).

Geometric properties
    Crack half-length, a = 0.1 in
    Plate half-width, b = 20 in (W in this text)
    Plate thickness, t = 1 in (B in this text)
    After selecting the "Add new stress range" button shown in Figure 7.12, the
    following are entered:
    Number of cycles: N = 3000; Tensile stress: σ_{max} = 15 ksi, σ_{min} = − 5 ksi
    Number of cycles: N = 7000; Tensile stress: σ_{max} = 3.5 ksi, σ_{min} = 0 ksi

Step 2 Material selection

Aluminum 7075-T6 is available in the software database and selected similarly
to the material selection steps described in Example 7.1. The following fracture
and FCG properties are listed for aluminum 7075-T6 in this software's database:

Fracture properties

K_{Ic} = 21,000 psi$\sqrt{in}$; plane-strain fracture toughness
A_k, B_k = 1; material constants that relate fracture toughness to thickness, see
Eqn. (2.16).

Fatigue crack growth properties
C_0 = 4.743e19 (in/cycle)/psi$\sqrt{in}$; crack growth rate intercept for R = 0
n = 3.5; crack growth rate slope
p = 0.5; NASGRO equation coefficient
q = 1; NASGRO equation coefficient
$\Delta K_{th.1}$ = 750 psi$\sqrt{in}$; threshold ΔK for $R \rightarrow 1$

Step 3 Calculation

Selecting "Calculate Results" (Figure 7.2) now generates the following: The number of cycles that can be achieved before failure is:
N_{fail} = 20,030 cycles
The total number of cycles in the stress history, N_{hist}, is 10,000 cycles. The number
of stress histories that can be repeated before failure is:

$$X_{hist} = \frac{N_{fail}}{N_{hist}} = 2 \text{ history repetitions to failure}$$

Based on LEFM, the panel will not fail before the next maintenance event.
However, this panel would never be allowed to stay in service if the crack were
found. The critical crack length occurs at 20,030 cycles; this is just slightly beyond
the number of cycles before the next maintenance event (20,000 cycles). The
calculated number of cycles to failure should always substantially exceed (on the
order of 6–8 times) the number of cycles between maintenance events.

7.3.5 REVERSING THE LOADING SEQUENCE

When the stress sequence is reversed N_{fail} = 27,030 cycles. That N_{fail} is not the same
when the loading sequence is reversed indicates that N_{fail} depends on stress history.
The plastic zone ahead of a crack tip may reduce or increase in size depending on the

FIGURE 7.11   Rectangular plate with semi-elliptical surface crack.

magnitude of $\Delta\sigma$. This change in plastic zone size causes residual stresses to develop; these along with phenomena such as crack blunting or crack closing will influence the crack growth response to the subsequent $\Delta\sigma$.

For example, if a single tensile load-unload is applied, the plastic zone may cause crack tip blunting. The plastic zone would decrease in size and compressive residual stresses would develop around it. If a compressive load is then applied, the blunted crack tip may become resharpened. Consider the reversed sequence; the compressive load might not completely close the crack face in the absence of residual compressive stresses. The subsequent tensile load-unload may result in a differently blunted crack tip that would respond differently (than the sharp crack tip in the first loading sequence) to any subsequent loading.

Example 7.6

A 5 cm thick rectangular plate made from HY-80 steel is simultaneously subjected to cyclical in-service tensile and bending stresses for 15,000 cycles. The maximum stress values are 270 MPa (tensile) and 70 MPa (bending). The R-ratio for both cases is 0.5. Determine the number of cycles until failure if a semi-elliptical surface crack exists as shown in Figure 7.11. The crack is 4 cm long and 1 cm deep. The plate is 1 m wide.

We will use the FCG calculator at mechanicalc.com for this problem.

INPUTS

The geometric and applied load inputs are entered as shown below.

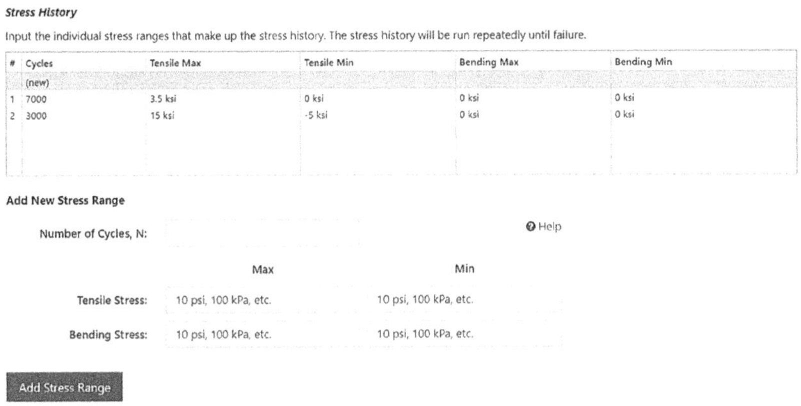

Stress History

Input the individual stress ranges that make up the stress history. The stress history will be run repeatedly until failure.

#	Cycles	Tensile Max	Tensile Min	Bending Max	Bending Min
	(new)				
1	7000	3.5 ksi	0 ksi	0 ksi	0 ksi
2	3000	15 ksi	-5 ksi	0 ksi	0 ksi

Add New Stress Range

Number of Cycles, N: ❷ Help

	Max	Min
Tensile Stress:	10 psi, 100 kPa, etc.	10 psi, 100 kPa, etc.
Bending Stress:	10 psi, 100 kPa, etc.	10 psi, 100 kPa, etc.

Add Stress Range

FIGURE 7.12   Load input fatigue crack growth calculator at Mechanicalc.com [1].

GEOMETRIC PROPERTIES

Crack Type: Elliptical surface crack
Crack depth, $a = 10\ mm$
Crack half-length, $c = 20\ mm$
Plate half-width, $b = 500\ mm$ (W in this text)
Plate thickness, $t = 50\ mm$ (B in this text)

MATERIAL SELECTION

HY-80 steel can be selected from the software's material database.

LOAD

Upon selecting the "Add stress range" button shown in Figure 7.12, the following can be entered:

Number of cycles, $N = 15000$
Tensile case, $\sigma_{max} = 270\ MPa$, $\sigma_{min} = 135\ MPa$
Bending case $\sigma_{max} = 70\ MPa$, $\sigma_{min} = 35\ MPa$

Note these are entered for the same case since the tensile and bending loads are simultaneous. The loading is NOT sequential.

CALCULATION

Selecting "Calculate Results" (Figure 7.2) now generates the following: The number of cycles that can be achieved before failure is:

$$N_{fail} = 93,502\ \text{cycles}$$

The total number of cycles in the stress history, N_{hist}, is 15,000 cycles. Therefore, the number of stress histories that can be repeated before failure is:

$$X_{hist} = \frac{N_{fail}}{N_{hist}} = 6.2\ \text{history repetitions to failure}$$

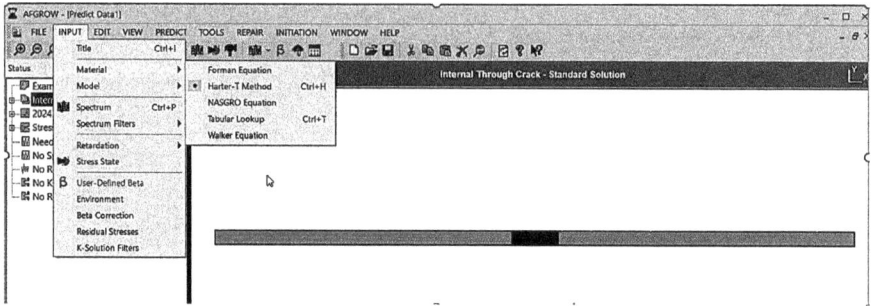

FIGURE 7.13   AFGROW 5.3.5.24 input options.

7.4   FATIGUE ANALYSIS WITH AFGROW

AFGROW is designed for actual cases that occur in industry, civilian and military. It therefore provides more options to handle the complexity of actual conditions. Some of these complexities include crack initiation, variable amplitude loading, residual stresses, corrosion material loss during crack growth, two cracks near each other, crack growth retardation, and time dependence. Figure 7.13 shows the material input options available for AFGROW software, version 5.3.5.24. Note that the material property definition starts with choosing which fatigue behavior model to use or the direct input of experimental data (tabular look up). For bonded composite repair analysis (beyond the scope of this text), the coefficient of thermal expansion and Poisson's ratio would also be needed. The dominant accepted da/dN vs. ΔK models are described below.

7.4.1   Crack Growth Model and Material Input

AFGROW provides the option of English units (i.e. stress (ksi), length (inch), force (kip), and temperature (Fahrenheit)) or SI (stress (MPa), length (m), force (MN), and temperature (Celsius)). The units are chosen by clicking on a small ruler at the bottom right corner on the main screen in Figure 7.13.

AFGROW's material input includes a choice of the da/dN vs. ΔK model, load and geometry combined. The other models available in addition to the NASGRO equation are the Walker equation, Forman equation, and the Harter T-method. These are described below. After choosing a da/dN vs. ΔK model (see Figure 7.13), the choice of material window will appear as shown in Figure 7.14. In this case, 7050-T74 aluminum is selected after choosing the Harter T-method crack growth model.

7.4.2   The Walker Equation

The Walker equation extends the Paris equation to account for the effect of the stress ratio R on the crack growth rate. The Walker equation is:

$$\frac{da}{dN} = C_0 \left(\Delta K \left(1 - R \right)^{m-1} \right)^n \qquad (7.10)$$

Material Database   ✕

ⓘ   The Harter T-Method is an adaptation of the Walker equation taken
     on a point-by-point basis. In this implementation 25 da/dN values are
     used to define the material behavior.
The method accounts for the stress ratio (R) shift independently at each da/dN
values to more accurately represent real crack growth rate data.
The data are contained in a user maintained library for several materials.

┌ Database File ──┐
│ ┌──┐   │
│ │ c:\program files\afgrow\matfile.md3   │   Browse ... │
│ └──┘   │
└──┘

┌ Selection ───┐
│ 7050-T74 PLATE   │
└───┘

Available Materials:

┌──┐
│ 2024 T-3 Bare Sheet LONG CRACK DATA   ∧ │
│ 2024-T851 - LONG CRACK   │
│ 6061-T6511 EXTRUSION   │
│ 7050-T74 PLATE   │
│ 7075-T6511 EXTRUSION   │
│ 7075-T73 L-T FORGING   │
│ 7075-T73 L-T [DRY AIR]   │
│ 7075-T73 T-L FORGING   │
│ Ti-6-4 ALPHA-BETA FORGING   ∨ │
└──┘

        ┌──────────────┐   ┌──────────────┐
        │   OK   │   │   Cancel   │
        └──────────────┘   └──────────────┘

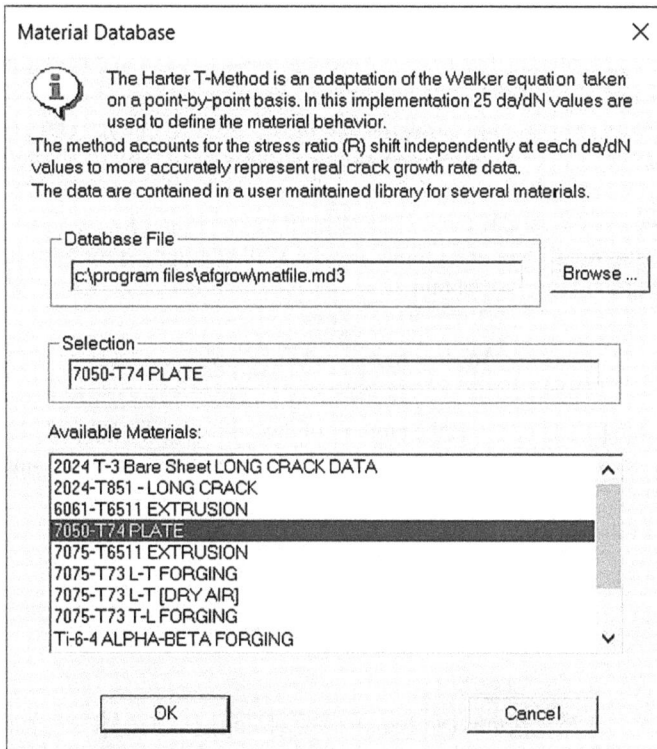

FIGURE 7.14 Material selection in AFGROW.

The effect of the stress ratio R on the crack growth rate is unique to each material. The parameter m is a measure of the magnitude of this effect and represents a shift in the da/dN vs. ΔK data as a function of the stress ratio R. C_0 is the value of the intercept constant C when $R = 0$.

7.4.3 THE FORMAN EQUATION

The Forman equation [4] was developed to more accurately model Region III of the da/dN vs. ΔK curve (See Figure 4.5) and accounts for the asymptotic behavior as the value of ΔK at fracture is approached. The form of the Forman equation used in AFGROW is:

$$\frac{da}{dN} = \frac{C(\Delta K)^n}{(1-R)K_c - \Delta K} \tag{7.11}$$

At high stress intensities (Region III), the plastic zone size becomes large relative to the crack size, so plasticity effects have a greater influence on the crack growth rate. Therefore, analysis using elastic-plastic fracture mechanics (EPFM) is recommended in addition to or in lieu of using the Forman equation. The Forman equation, however, can also be used to determine the mean stress effects in Region II.

7.4.4 THE HARTER T-METHOD

Tabular crack growth rate data may be available; however, it could be within a limited set of R-values. If the user has done their own experiments or has access to external data from recent experiments, these may be input into the system. If not, one alternative is to use the Walker equation on a point-by-point basis to extrapolate/interpolate data for any R-value. This approach is known as the Harter T-method [5]. Suppose there are two da/dN vs. ΔK curves given by R-ratios R_1 and R_2, as shown in Figure 7.15. Assuming $R_1, R_2 > 0$, and employing the Walker equation for both R-ratios at a given crack growth rate, we obtain

$$\Delta K_1 \left(1 - R_1\right)^{m-1} = \Delta K_2 \left(1 - R_2\right)^{m-1} \tag{7.12}$$

The m value $[0 - 1]$ controls the amount of data shifting. As m increases, the shifting decreases. Note m is NOT a slope, a distance, or stress intensity and it has no dimensions.

Solving Eqn. (7.12) for m yields

$$R_1, R_2 > 0, m = 1 + \left[\log_{10}\left(\frac{\Delta K_1}{\Delta K_2}\right) \middle/ \log_{10}\left(\frac{1 - R_2}{1 - R_1}\right) \right] \tag{7.13}$$

$$R_1 < 0, R_2 \geq 0, m = 1 + \left[\log_{10}\left(\frac{K_{max,1}}{\Delta K_2}\right) \middle/ \log_{10}\left(\left(1 - R_1\right)\left(1 - R_2\right)\right) \right] \tag{7.14}$$

$$R_1, R_2 < 0, m = 1 + \left[\log_{10}\left(\frac{K_{max,1}}{K_{max,2}}\right) \middle/ \log_{10}\left(\frac{1 - R_2}{1 - R_1}\right) \right] \tag{7.15}$$

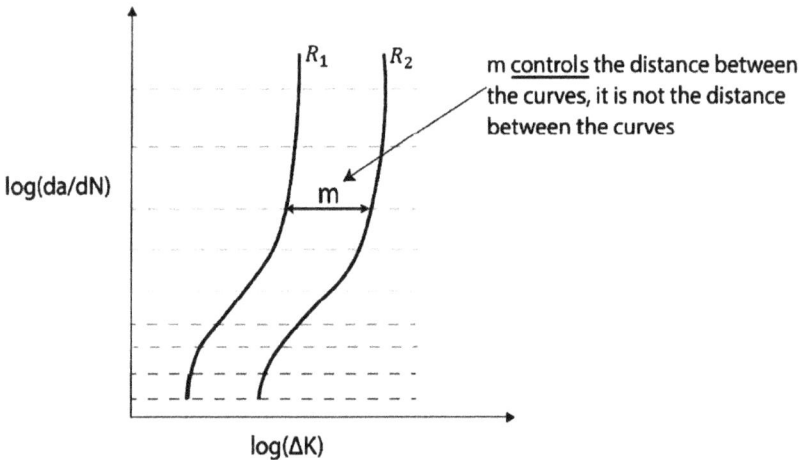

FIGURE 7.15   Harter T-method.

Concept Check 7.1

The Harter T-method can be applied if you need to develop a *da/dN* curve at an *R*-value that falls between two *R*-values for which you have *da/dN* data. What might you do if you have *da/dN* curves for multiple R-values?

7.4.5 ADVANCED INPUTS

The more advanced inputs offered by AFGROW include (i) environmental data, (ii) a beta correction factor, (iii) residual stress, and (iv) a K-solution filter.

- Environment: If environment-based crack growth rate data is available, this option allows the user to input this data.
- Beta correction: The software provides a method (based on Gaussian integration) to estimate stress intensity factors for cases which may not be an exact match for one of the stress intensity solutions in the library. The details of this method are beyond the scope of this text.
- Residual stress: Not to be confused with the residual strength (curve), this option provides an opportunity to input existing residual stresses. The associated intensities are added at user defined crack length increments.
- K-solution filter: This option allows stress intensity solutions for the axial, bending, and/or bearing load case to be modified independently of the applied tensile and/or compressive spectrum loading.

7.4.6 MODEL GEOMETRY AND LOAD

Figure 7.16 shows various model options in the AFGROW software; in this scenario, the "Internal Through Crack" in a thin plate option is selected. Note the tabs that allow for dimensions and loading input. Loading input may include axial, bending, or bearing stress fraction values.

7.4.7 LOAD INPUT

The load (or stress) input is selected from the load tab shown in Figure 7.17. The stress fraction is the ratio of the specific applied stress value (tensile, bearing, bending) to some reference stress. The reference value can be chosen to be the maximum (absolute value) applied stress. The geometrical input process is straightforward and involves the direct input of the geometrical parameters.

7.4.8 STRESS STATE

AFGROW allows the user to specify whether a plane stress or plane strain analysis is required. This is shown in Figure 7.18. It may also use the material properties and the applied loads to determine the stress state automatically.

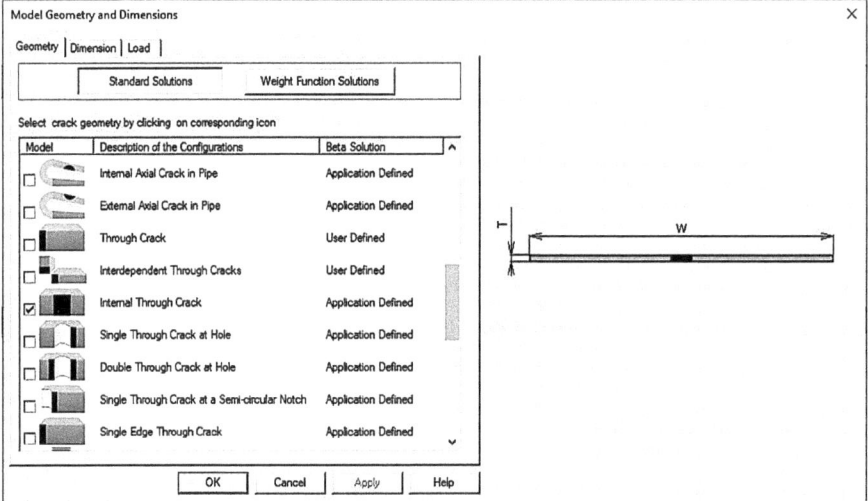

FIGURE 7.16   Various crack types within different geometries.

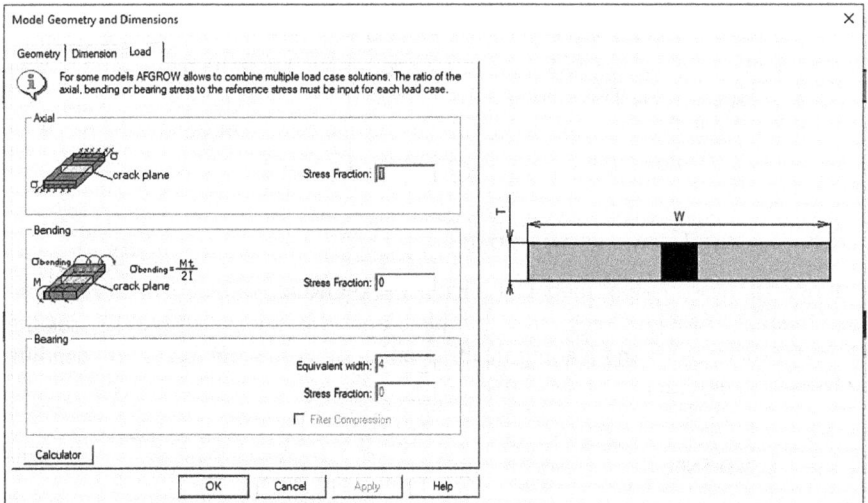

FIGURE 7.17   AFGROW load input.

7.4.9   STRESS SPECTRUM INPUT

Since AFGROW is designed for real world solutions, it requests input from a spectrum file. The stress spectrum is imported by selecting the "spectrum" option from the "input" option in the top menu (Figure 7.13). This spectrum option is shown in Figure 7.19. Note there is a choice to use constant amplitude data, create your own spectrum file, or import an existing spectrum file.

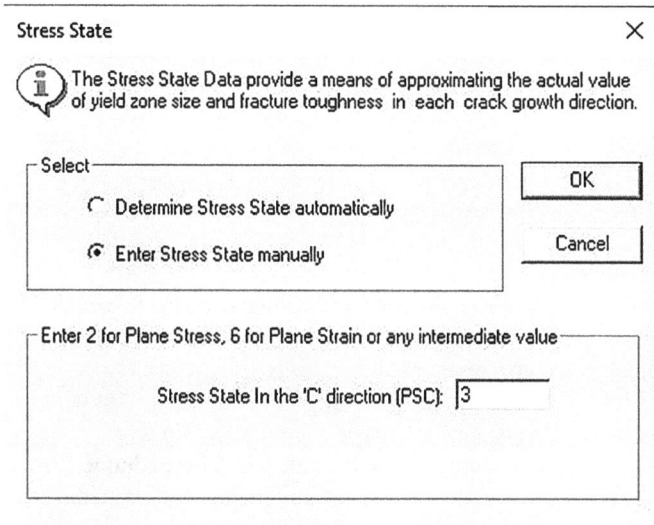

FIGURE 7.18 Stress state option.

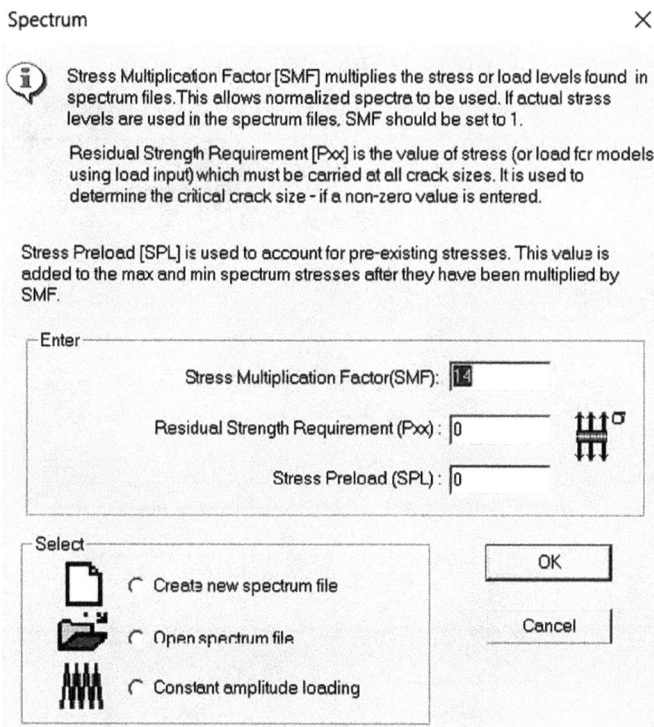

FIGURE 7.19 AFGROW stress spectrum request.

If the spectrum data are normalized (i.e. all values are divided by some maximum reference stress value so their new values now lie between 0 and 1), the user then chooses a factor to predict life at different stress levels for that spectrum. AFGROW refers to this factor as the spectrum multiplication factor (SMF).

In some cases, where the maximum stress (absolute value) that occurs in the spectra is unknown, the user may want to continuously check for failure. The residual strength option allows the user to provide a stress value whose structure must be able to withstand all crack sizes. If the residual strength value is set to zero, failure occurs based on the current applied stress.

Stress preload (SPL) accounts for pre-existing stresses. It is added to the maximum and minimum stresses (multiplied by SMF) in the analysis.

When the constant amplitude loading is selected, the user has the option to enter the R-ratio and a block size. The block size sets the number of cycles over which each pass of integration occurs. Increasing block size will tend to reduce the analysis time; however, the accuracy may be reduced. The dialogue box is shown in Figure 7.20. The time dependent option allows the user to relate the block to a time interval, so the program will also output time to failure, as well as the number of cycles.

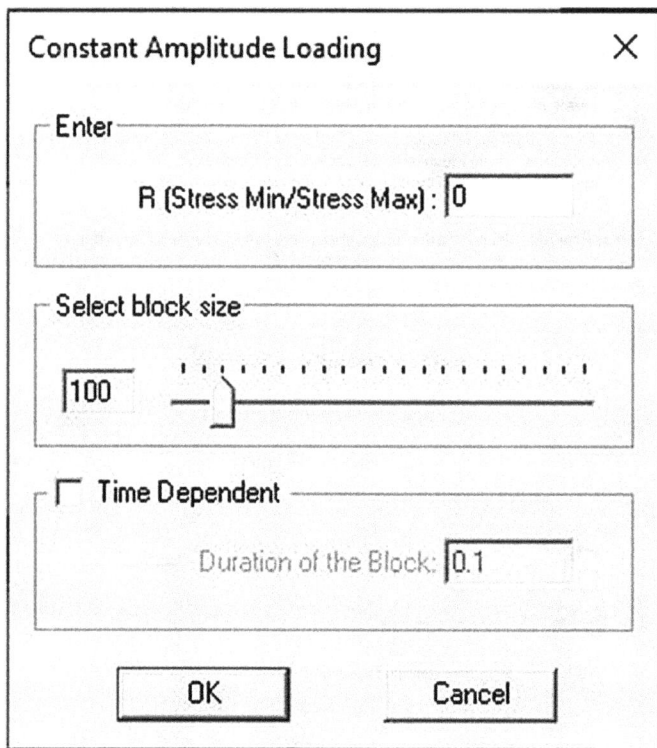

FIGURE 7.20 Constant amplitude dialogue box.

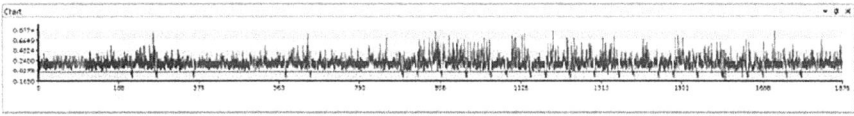

FIGURE 7.21   Variable amplitude spectrum.

Variable spectrum files can be imported or created. An example of a realistic variable amplitude spectrum is shown in Figure 7.21.

7.4.10   CALCULATION

The calculation is now performed by selecting "predict" from the main options (Figure 7.13) and selecting the "run" option from the drop-down menu. The software can provide the output in various forms (Excel spreadsheet, data file, plot file); the screen output is shown in Figure 7.22. Iterative crack growth data are also included in the output; however, these are not shown.

```
Crack Growth Model and Spectrum Information
Title: Example Problem

Load: Axial Stress Fraction: 1, Bending Stress Fraction: 0, Bearing Stress Fraction: 0
Crack Model: 2010 - Internal Through Crack - Standard Solution
Initial surface crack length      (C):   0.1500
Thickness   :    0.100
Width       :    4.000

Young's Modulus =10400, Poisson's Ratio =0.33 , Coeff. of Thermal Expan. =1.34e-005

No crack growth retardation is being considered

Determine Stress State automatically (2 = Plane stress, 6 = Plane strain)

No K-Solution Filters

Harter T-Method crack growth rate approach is being used
 For Reff < 0.0, Kmax is used in place of Delta K
 Material: 7050-T74 PLATE
 Lower 'R' value boundary: -0.33
 Upper 'R' value boundary: 0.8
 Plane strain fracture toughness: 33
 Yield stress: 65

Failure is based on the current load in the applied spectrum

Vroman integration at 5% crack length

Spectrum: Constant amplitude loading
Spectrum multiplication factor:   10.000
SPL:    0.000
The spectrum will be repeated up to 999999 times
Total Cycles: 400
Levels: 1
Subspectra: 1
Max Value: 1
Min Value: 0.5

No Spectrum Filters

Critical Crack Length is Based on the Maximum Spectrum Stress
Critical crack size in 'C' direction=1.8425, Stress State=2 (Based on Kmax criteria)
Critical crack size in 'C' direction=1.62482, Stress State=2 (Based on Net Section Yield criteria)
```

FIGURE 7.22   AFGROW output.

Example 7.7   (AFGROW)

An Al 7050-T74 plate is 0.15 in. thick (T) by 1.5 in. wide and is fastened to another component via a pin of diameter, d, 0.3 in. When in service, a remote cyclical loading (constant amplitude, $R = 0.5$) results in a bypass stress $\sigma_{by} = 12$ ksi and a bending stress $\sigma_{bd} = 7$ ksi. A maximum pin load of 675 lbs is also transferred via the hole-fastener interface. The plate under loading is shown in Figure 7.23. Calculate the crack growth life for a crack of length, c, 0.05 in. and depth, a, 0.07 in. located at the corner of the hole. Determine the number of cycles to failure. Assume no crack retardation occurs.

Note AFGROW uses W for total width (as opposed to $2W$).

The remotely applied gross stress may be used as a reference stress to define the stress fractions. Let σ_T represent the remotely applied gross stress and σ_p represent the stress on the cross-section due to the pin load

$$\sigma_T = \sigma_{by} + \sigma_p + \sigma_{bd}$$

$$\sigma_T = 12 + \frac{\left(675 \times 10^{-3}\right)}{\left(1.5 \times 0.15\right)} + 7 = 22 \text{ ksi}$$

The axial stress fraction is: $12/22 = 0.54$
The bending stress fraction is: $7/22 = 0.32$
The bearing stress fraction is: $\dfrac{\left(675 \times 10^{-3}\right)}{\left(0.3 \times 0.15\right)}/22 = 0.68$

The K_{Ic} solution will then be given by:

$$K_{Ic} = \left(0.54 Y_{ax} + 0.32 Y_{bd} + 0.68 Y_{br}\right)\sigma_T\sqrt{\pi a}$$

SOLUTION

Step 1 Units
Open AFGROW and ensure that the English units are selected; this option is located at the bottom right corner of the screen. This is shown in Figure 7.24. AFGROW's English units used in FM analysis are stress (ksi), length (inch), force (Kip), and temperature (degrees Fahrenheit).

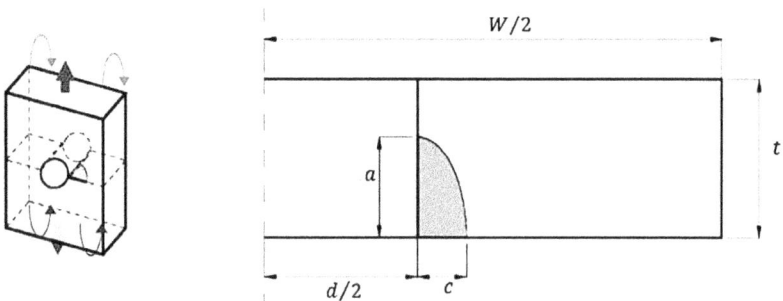

FIGURE 7.23   Plate with crack at fastener hole for Example 7.7.

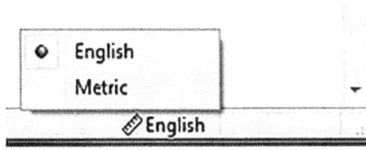

FIGURE 7.24 AFGROW units.

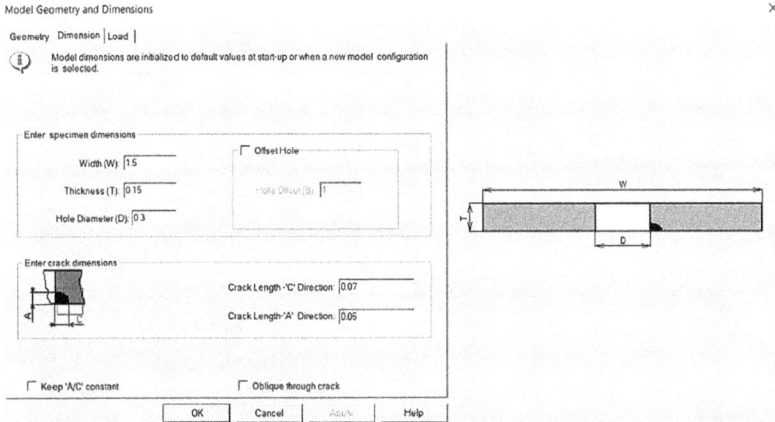

FIGURE 7.25 Geometry input.

Step 2 Model input
Select "Input" from the top menu, then "model" followed by "classic model" in the dropdown menu (see Figure 7.13). Select the "Single Corner Crack at Hole." This selection window is already shown in Figure 7.16.

Step 3 Material input
Navigate to the material input as previously described in Section 7.4.1. Select the Harter T-method and Al 7050-T74 (see Figure 7.13).

Step 4 Geometry and load input
The geometrical properties are entered directly after selecting the dimension tab as shown in Figure 7.25. The stress fractions are entered directly after selecting the load tab as shown in Figure 7.26.

Step 5 Spectrum input
Navigate to the spectrum option as described previously for Figure 7.20. Choose the SMF to be 22 (gross remote stress) and the R-ratio to be 0.5, and set the number of the block size to 200.

Step 6 Calculation
Select "predict" from the main menu options (Figure 7.13) and then select the "run" option from the drop-down menu. The software predicts that failure occurs after 24,470 cycles when the crack size in the "C" direction is 0.57 inches. The software automatically determined a plane stress condition was suitable (stress state 2).

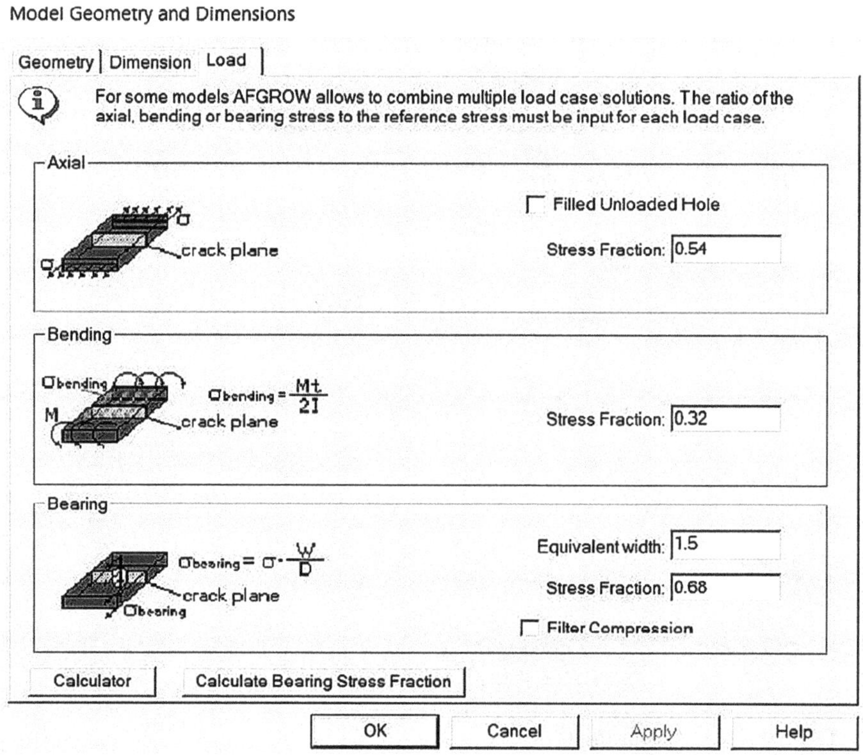

Model Geometry and Dimensions

Geometry | Dimension | Load |

For some models AFGROW allows to combine multiple load case solutions. The ratio of the axial, bending or bearing stress to the reference stress must be input for each load case.

Axial

crack plane

☐ Filled Unloaded Hole

Stress Fraction: 0.54

Bending

$\sigma_{bending}$

$\sigma_{bending} = \frac{Mt}{2I}$

crack plane

M

Stress Fraction: 0.32

Bearing

$\sigma_{bearing} = \sigma \cdot \frac{W}{D}$

crack plane

$\sigma_{bearing}$

Equivalent width: 1.5

Stress Fraction: 0.68

☐ Filter Compression

| Calculator | Calculate Bearing Stress Fraction |

| OK | Cancel | Apply | Help |

FIGURE 7.26   Stress fraction input for AFGROW.

7.5 SUMMARY

The approaches to solving elementary static and fatigue LEFM problems using software have been presented. Key concepts include the superposition of solutions (e.g. tensile and bending) that result in the same crack loading mode, and checking whether plane stress or plane strain applies; most software will do these automatically. However, standards evolve, and it is safest to check what formulas are being applied by the chosen software. Factors of Safety may be determined by directly comparing calculated stress intensity values to the material fracture toughness values in the stress intensity equation and/or by using the residual strength curve. In fatigue analysis, it is important to know where on the $\frac{da}{dN}$ vs. ΔK curve the crack growth model is being applied, and what R-ratio is being used (constant amplitude), as these influence the choice of c and rack growth model. Other factors that should be researched for both the static and fatigue cases include the existence of residual stresses, the possibility of crack retardation, environmental effects, and proximity to other cracks or edges.

PROBLEMS

7.1. The stress fractions in Example 7.7 do not sum to 1. Shouldn't they necessarily sum to 1? Why? Why not?

7.2. For a certain FCG problem, the crack length is already in Region III of the $\frac{da}{dN}$ vs. ΔK curve. Which model is preferable? Paris, Walker, or Forman? Why?

7.3. Tabular growth rates at different R-ratios are available for a certain material. However, for a given FCG problem with this material, the growth rate data for the R-ratio specific to this scenario is not available. Which crack growth model would be suitable to apply here? Why?

7.4. Research and discuss a failure case study where the environment played a major role in the growth of a crack leading to unexpected failure. Discuss how consistent low temperatures, a saline environment, or high humidity might affect the response of a crack when subjected to service loads.

7.5. Research and describe a case study where residual stresses had a significant effect on crack growth. Where did these residual stresses come from? How can they be used in a beneficial way? If not, how can they be minimized? Describe how software might account for residual stresses in the crack growth analysis.

7.6. List and describe what additional inputs or approaches are needed for variable amplitude loading compared to constant amplitude loading.

Write a program in a mathematical programming language, use a spreadsheet or a relevant software package to solve problems 7.7 and 7.8.

7.7. A center cracked plate with fracture toughness of 37 MPa$\sqrt{m}$ contains an initial crack of length $2a = 20$ mm. It is subjected to constant amplitude cyclic tensile loading which results in an R-ratio of 0.3 and a stress amplitude of 14 MPa. Assuming the fatigue crack growth rate a is governed by the equation $da/dN = 0.48 \times 10^{-11}(\Delta K)^3$, calculate the crack growth rate when the crack length has the following values: 35 mm, 45mm. Estimate the number of cycles to failure.

7.8. A 4 cm thick rectangular plate made from HY-80 steel is subjected to cyclical bending stresses for 25,000 cycles. The maximum stress amplitude is 50 MPa with an R-ratio of 0.4. Determine the number of cycles until failure if an elliptical surface crack exists as shown in Figure 7.12. The crack is 3 cm long and 1 cm deep. The plate is 1 m wide.

REFERENCES

[1] A. M. Kirby, MechaniCalc Inc, Fracture Mechanics Calculator www.mechanicalc.com, MechaniCalc, Inc.

[2] V. Lawrence and R. Forman, "Structure and Applications of the NASA Fracture Mechanics Database," in *Computerization and Networking of Materials Databases: Third Volume*, West Conshoocken, PA, ASTM International, 1992, pp. 370–381.

[3] J. Newman Jr., "Effect of Constraint on Crack Growth Under Aircraft Spectrum Loading," NASA Tecnical Memorandum 107677, Hampton, VA, 1992.

[4] R. Forman, V. Kearney and R. Engle, "Numerical Analysis of Crack Propagation in Cyclic-Loaded Structures," *Journal of Basic Engineering*, vol. 89, no. 3, pp. 459–463, 1967.

[5] J. Harter, "Damage tolerance management of the x-29 vertical tail," in *USAF Structural Intergity Conference*, San Antonio, 1991.

8 Finite Element Method Use in Fracture Mechanics

OBJECTIVES

After studying this chapter, the student should be able to:

1. Know the typical computer aided engineering (CAE) terminology used to model fracture mechanics (FM) problems with the finite element method (FEM).
2. Understand the structure and processes needed for modeling crack behavior with FEM tools.
3. Know the various options for modeling cracks and crack propagation using FEM.
4. Be able to solve line integral problems by creating their own code for mathematical software and by hand.
5. Understand the input and processes needed to apply the extended finite element method (XFEM) to FM problems.
6. Know the differences between FEM and XFEM input and capabilities in modeling crack behavior.

8.1 AN INTRODUCTION TO THE FINITE ELEMENT METHOD (FEM)

Closed-form methods of solving engineering problems are typically limited to simple geometries such as beams, cylinders, spheres, and plates; numerical methods are usually needed for complex geometries. A numerical method usually discretizes the problem into simpler geometries and the local solution to the simpler geometry is then determined. The global solution is then found by combining the local solutions and imposing boundary conditions. Local solutions are usually combined by applying certain continuity relationships at the boundaries between the simpler geometries. The trapezoidal rule, used to find the area under a curve, is an example of a numerical method.

A numerical method finds an approximate solution. Accuracy is increased by increasing the level of discretization. As discretization increases, the solution will converge to a unique value. Convergence does not necessarily mean the solution is correct as the mathematical algorithm may have generated local maxima or minima. It is therefore important to understand the algorithmic assumptions and to compare numerical solutions with other feasible solutions. The latter may be in the form of

DOI: 10.1201/9781003052050-8

closed-form solutions for simpler geometries, experimental results, historical data, or results from other numerical methods.

FEM or finite element analysis (FEA) is a numerical method used in solid mechanics to determine field variables such as stresses and strains within a given geometry, subjected to specific load applications and displacement constraints. In FEM, the simple geometries or shapes are called *elements*; the points on the elements at which field variables (e.g. stress, strain) are calculated are called *nodes*. Values of the field variable at locations other than the nodes are determined by interpolation.

Figure 8.1(a) shows various 2D and 3D elements while Figure 8.1(b) shows an FE model of a complex geometry. Note that nodes may be defined at locations other than the vertices. When the nodes are located at the vertices only, the elements are called first-order elements; field values at other locations in the element are approximated by using a linear function. When nodes are also located at the midpoints, the elements are second-order elements; field values at other locations in the element are approximated using a quadratic function. The choice of element is very important in FE analysis. Factors to consider in element selection include: the element geometry, element dimensionality (1D, 2D, 3D), element order (first order, second order), the type of analysis (static, dynamic, linear, nonlinear), and run time.

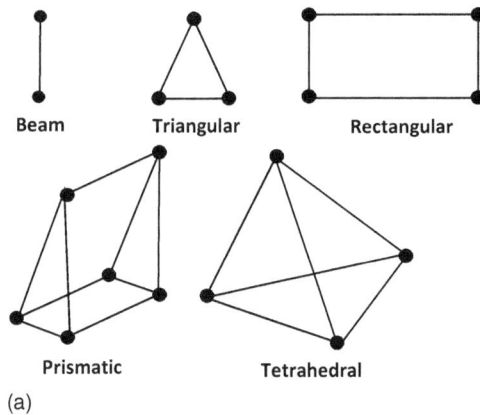

(a)

(b)

FIGURE 8.1   (a) Various 2D and 3D elements. (b) FEM model of a quadcopter propeller.

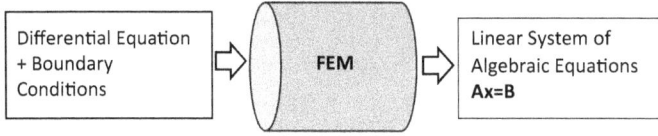

FIGURE 8.2   Schematic representation of the FEM process.

Most engineering problems related to structural response are modeled using differential equations. The validity of the equations is typically limited to a certain geometry or bounded volume, therefore the conditions at the boundary are unique to each problem. The constraints at the boundary may be due to a variety of physical phenomena such as loading, displacement, temperature, pressure, and magnetic field strength. The differential equation with boundary conditions is referred to as a *boundary value problem*. FEM converts the boundary value problem to an algebraic problem, as shown in Figure 8.2. Typical FEM inputs are loads and displacement constraints; typical outputs will include stresses, strains, and displacements.

The FEM process for stress analysis typically requires the following steps:

- Define the type of study (e.g. static, dynamic, thermal);
- Define the geometry;
- Input material properties (e.g. E, ν);
- Apply mesh to geometry;
- Apply loads/boundary conditions;
- Solve for field variables such as stresses, strains, and displacements.

8.2   FEM FOR FRACTURE MECHANICS

FEM is advantageous over the previously discussed methods in Chapter 7 in that complex geometries can be fully modeled; stresses, strains, and displacements can be determined everywhere within the model. Additionally, a wider variety of crack configurations, loadings, and crack interactions can be studied with FEM. FEM crack propagation analysis usually requires the definition of distinct initially bonded contact surfaces between which the crack will propagate. Cracks are modeled by allowing overlapping nodes and then applying node separation criteria or using contour integrals. Contour integrals are also used to predict the onset of cracking in 2D or 3D problems. Recall that stress intensity may be determined if the J-integral is known.

The FEM process for crack modeling typically involves the FEM steps for stress analysis and additional steps to define the crack on the relevant face of the part and a method for evaluating the field variables in the vicinity of the crack. These definitions are not trivial; theoretically there is a singularity at the crack tip! Special care must be taken over the decision regarding model choice and input. Because of the complexity at the crack tip, the area in the vicinity of the crack will typically need a finer mesh relative to the rest of the part.

8.3   DEFINITIONS AND TERMINOLOGY

The terminology and methods for modeling cracks are similar across various platforms, therefore we will review the common terminology and approaches. Protocol for ABAQUS [1] is used here as an example. The vocabulary must be specific when using all CAE software.

- A native part: An original part.
- Part instance: A copy of the original part. Part instances are typically used in an assembly module.
- Orphan mesh part: A part defined only by nodes, elements, and surface sets. It does not contain any feature information, and geometric features cannot be added. Crack Front: The crack front is defined by selecting certain features (e.g. vertices, lines) in the region at the forward end of the crack. A crack front for a 2D native part may be defined using a single vertex, connected edges, or connected faces.

The 2D crack front definition options are shown in Figure 8.3. In the case of an orphan mesh, the vertex, connected edges, or connected faces would be replaced by a node, a set of connected element edges, or connected elements respectively, as shown in Figure 8.4.

A crack front for a 3D native part may be defined using connected edges, connected faces, or connected cells. These options are shown in Figure 8.5. In the case of an orphan mesh, the line, face, or cell would be replaced by element edges, element faces, or a set of elements.

8.3.1   CRACK TIP/LINE SELECTION

For the 2D case, the crack tip is defined by selecting a vertex or a node. If a vertex or a node is used to define the crack front, the same vertex or node will then define the crack tip. For the 3D case, the crack line is defined by selecting edges or element

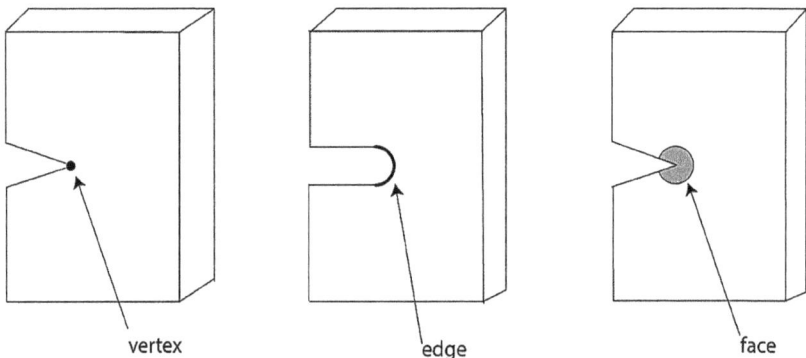

vertex   edge   face

FIGURE 8.3   Selection options for a 2D crack front definition for a native part (Image Source: [1]).

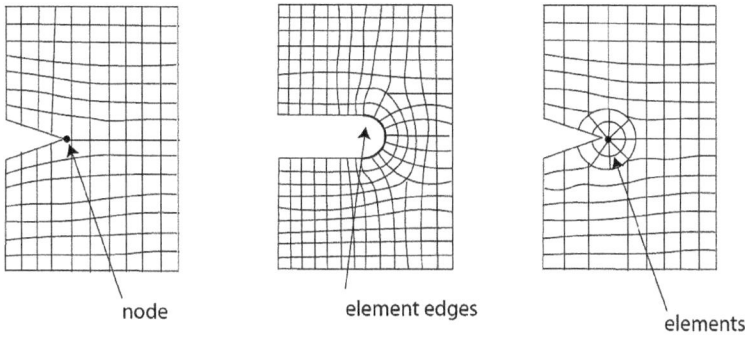

FIGURE 8.4 Selection options for a 2D crack front definition for an orphan mesh (Image Source: [1]).

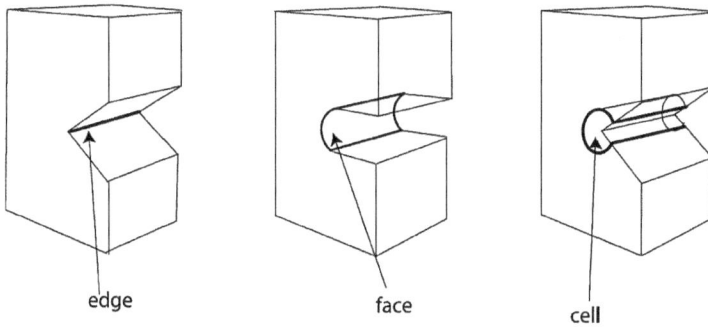

FIGURE 8.5 Selection options for a 3D crack front definition for a native part (Image Source: [1]).

edges that form a continuous line. If a selected edge or element edge is used to define the crack front, the same edge will then define the crack line.

8.3.2 CRACK EXTENSION DIRECTION

To define the crack extension direction, either the start and end points of a crack extension direction vector(s) at the crack tip (or along the crack line) or the start and end points of the normal to the crack plane will need to be specified. Selecting the latter option would imply the crack is a flat plane (so it is defined by a single normal).

8.3.3 CRACK SURFACES

Potential crack surfaces are typically modeled as contact surfaces. Some initial condition must be defined to identify which part of the crack is initially bonded. A node set may then be specified.

8.3.4 VIRTUAL CRACK CLOSURE TECHNIQUES (VCCT)

Energy is required to both open and close a crack. VCCT assumes that the strain energy released when a crack propagates by a certain amount is equal to the energy required to close the crack through the same distance. VCCT is sometimes used to characterize the energy release rate (G) for non-isotropic materials (e.g. delamination of composite structures). The enhanced VCCT allows the user to control the onset and growth of a crack using two different critical fracture energy release rates.

8.3.5   CRACK PROPAGATION

Crack propagation analysis is typically carried out on a nodal basis. The crack-tip node debonds when some fracture criterion is achieved. In ABAQUS, the fracture criteria appropriate for linear elastic fracture mechanics (LEFM) are critical stress, VCCT, or enhanced VCCT.

8.4   FEM AND FATIGUE CRACK GROWTH (FCG)

For structures subjected to sub-critical cyclic loading, a FEM-based FCG model may be developed. This is a quasi-static analysis characterized by the fracture energy release rate; it uses a classical incremental method for each loading cycle. A FEM-based FCG analysis also offers the ability to incorporate thermal loads and account for the change of contact conditions and geometric nonlinearity, and to model progressive delamination growth for laminated composites.

FEM crack propagation analysis can determine the following:

* Critical stress at a certain distance ahead of the crack tip;
* Critical crack opening displacement;
* Crack length versus time;
* Onset and growth of a crack (VCCT, enhanced VCCT);
* Fatigue crack growth criterion based on the Paris law.

8.5   EXTENDED FINITE ELEMENT METHOD (XFEM)

XFEM is used to study the onset and propagation of cracking in quasi-static problems or to simulate the fracture and failure of a structure under high-speed impact in an implicit dynamic analysis. It may not be available for all FE models (e.g. shell vs. solid model, 2D vs. 3D). XFEM allows the user to model crack growth along an arbitrary, solution-dependent path without the need to remesh their model.

XFEM is an extension of conventional FEM, based on the concept of the partition of unity. XFEM defines an "enrichment" area around a crack tip and in regions where the crack tip might grow. A finer mesh is created by splitting the special volume elements in the enrichment zone from the center of the element. The enrichment regions in XFEM are, however, very computationally expensive. Simulations become slower as the enrichment area increases. XFEM is therefore not easy to scale up to large projects. The results may also be influenced by the underlying mesh [2].

XFEM requires the specification of a *crack domain* and a *crack growth solution dependent path* or that it is stationary. An *initial crack location* may also be defined, or the software may search for regions that are experiencing principal stresses and/or strains greater than the maximum damage values specified by traction-separation laws.

TABLE 8.1

Comparison of FEM and XFEM for FM Applications

	FEM	XFEM
Mesh refinement	Needs refined mesh around crack tip	May use a coarse mesh
Crack geometry	Must conform	Not necessarily conform
Crack propagation	Remesh after each step	No remeshing needed
Damage model	Yes, debonding of nodes	Yes, damage
Crack initiation	Yes	Yes
2D	Yes	Yes
3D	Yes	Yes, limited

The crack must be defined as a separate part, the parts will then need to be assembled. The differences between XFEM and FEM for FM applications are provided in Table 8.1.

8.6 LINE INTEGRAL CALCULATION

The J-integral is a line integral. It is useful to understand how a line integral may be calculated in general for these reasons:

- It provides a foundation for the understanding of the coding behind domain and interaction integrals used by the software;
- Some software packages may allow the user to input additional code to address unique FM scenarios;
- The engineer may be part of a team tasked with developing FM software for problems unique to a particular industry's FM needs;
- The engineer may have to develop solutions using more general mathematical software (Excel, PYTHON, MATLAB) if specialized FM software is unavailable.

Example 8.1

Figure 8.6 illustrates a thin plate with an edge crack; by modeling via FEM the plate has been discretized into quadrilateral elements as shown. The traction vector along the contour Γ is predicted by the FE model to be zero on the vertical faces (BC, DE, FA) and crack surfaces Λ; $10\hat{i}$ N/m² on the top face, CD; and $-10\hat{i}$ N/m² on the bottom face, AB.

The displacements at all points along Γ can be approximated by the displacement vector $\bar{u} = 5xy\hat{i} + 3x^2\hat{j}$. Assuming the strain energy, U, is negligible, determine the J-integral around the contour Γ.

Since U is negligible, the J-integral is given by

$$J = \int_{\Gamma} \left(-\vec{T}.\frac{\partial \bar{u}}{\partial x} \right) ds$$

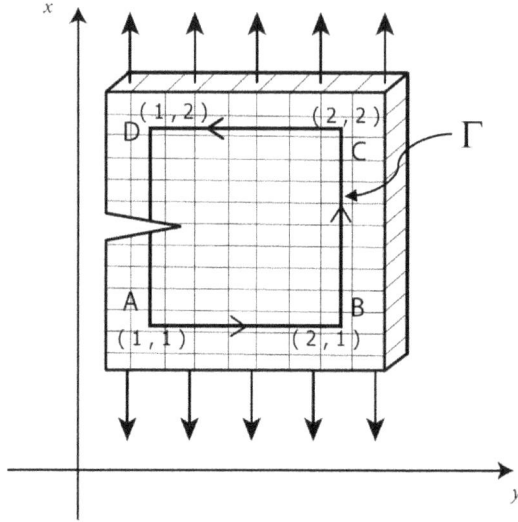

FIGURE 8.6   J-Integral for an Edge Crack in a Plate.

Along AB, $ds = dx$

$$J_{AB} = \int_1^2 \left(-\left(-10\hat{i}\right)\cdot\left(5y\hat{i} + 6x\hat{j}\right)\right)$$

$$= \int_0^1 50y\,dx$$

Since $y = 1$ along AB,

$$J_{AB} = \int_1^2 50\,dx = 50\,N/m$$

Along BC, $ds = dy$; however, since $\vec{T} = 0$, $J_{BC} = 0$.
   Along CD, $ds = -dx$

$$J_{BC} = \int_2^1 \left(\left(-10\hat{i}\right)\cdot\left(5y\hat{i} + 6x\hat{j}\right)\right)(-dx)$$

$$= \int_2^1 50y\,dx$$

Since along $y = 3$ CD

$$J_{CD} = \int_2^1 150\,dx = -150\,N/m$$

Along DE, FA, $ds = -dy$; however since $\bar{T} = 0$, $J_{BC} = 0$.
Along crack surfaces Λ, $\bar{T} = 0$, therefore $J_\Lambda = 0$.
Therefore

$$J = J_{AB} + J_{BC} + J_{CD} + J_{DA} + J_\Lambda$$
$$= 50 + (-150)$$
$$\boxed{= -100\,N/m}$$

Note the displacement vector was provided in the above example. In general, the displacements and corresponding displacement gradients (i.e. $\frac{\partial \bar{u}}{\partial x}$) would have been approximated by the FE software.

If the strain energy, traction, and displacement vector fields along the contour are known, a code to determine the J-integral may be written using any language. Example 8.2 provides an example of performing a line integral using mathematical software.

Example 8.2

The MATLAB 2019a code shown in Figure 8.8 determines the line integral along a curve $y = f(x)$ between specific x-value limits for some function $F = g(x, y)$. Note future MATLAB codes may slightly deviate from what is shown here.

Compute the integral $\int x^2 y\,ds$ along the line $y = x$, for $x = 0$ to $x = 2$, (a) manually, (b) using MATLAB, and (c) using MATHEMATICA.

(a) For $y = x$,

$$\frac{dy}{dx} = 1$$

Let ds represent incremental distance along the line $y = x$

$$ds = \sqrt{1 + \left(\frac{dy}{dx}\right)^2} = \sqrt{1 + 1^2}$$

$$\int x^2 y\,ds = \int_0^2 x^2 * x\sqrt{1 + 1^2}\,dx = \sqrt{2}\int_0^2 x^3\,dx = \sqrt{2}\left[\left(\frac{x^4}{4}\right)\right]_0^2 = 4\sqrt{2} = 5.6569$$

(b) If entered in the MATLAB code shown in Figure 8.7, the following would be entered: $g(x, y) = y * x^2$, $F = x$ and x_limits = $[0, 2]$
(c) The same line integral performed in MATHEMATICA could be done using the script in Figure 8.8.

```
syms x y %syms var1 ... varN creates symbolic scalar variables var1 ... varN
y=input("enter the curve f(x) along which line integral will occur:' ");
F=input("enter the function to be integrated, f(x,y):' ");
disp(y);
disp(F);
x_limits=input("enter the lower and upper limits for x as a vector:");
dy_dx=diff(y,x);%computes the derivative dy/dx
disp(dy_dx);
ds=sqrt(1+(dy_dx)^2);
disp(ds);
integrand=f*ds;%defines integrand
line_int=int(integrand, x_limits(1,1), x_limits(1,2));%reads limits from x_limits input
disp(line_int);
disp(eval(line_int));
```

FIGURE 8.7 MATLAB script to determine the line integral.

$$y = x;$$
$$F = x^2 y;$$
$$dydx = D[y,x];$$
$$ds = \sqrt{1 + (dydx)^2};$$
$$integrand = F * ds;$$
$$lineintegral = \int_0^2 integrand\ dx$$

FIGURE 8.8 MATHEMATICA script to solve the line integral in Example 8.1.

8.7 J-INTEGRAL AND FEM

Recall the J-integral discussed in Section 3.4, which was defined as the negative of the rate of change of the potential energy, Π, of deformation for an infinitesimal crack growth in a nonlinear elastic body with respect to crack growth area, A. In other words, it defines the amount of energy released per unit area of crack surface increase.

$$J = -\frac{\partial \Pi}{\partial A} \tag{8.1}$$

Further analysis yielded

$$J = \int_\Gamma \left(U dy - \vec{T} . \frac{\partial \vec{u}}{\partial x} ds \right) \tag{8.2}$$

where Γ is any contour surrounding the crack tip, U is the strain energy density, $\vec{T}$ is the traction on Γ, and $\bar{u}$ is the displacement on an element along arc length s (see Figure 3.13). The strain energy density U was given by Eqn. (3.17), i.e. $U = \int_0^\varepsilon \sigma\, d\varepsilon$.

In the determination of the J-integral, the following criteria must be satisfied:

• The crack is not loaded on its faces;
• The path direction is counterclockwise;
• The initial and end points lie on the two crack faces;
• The J-integral is a path independent line integral.

Concept Challenge 8.1

Software can calculate successive J-integrals for the same crack. Is there a significant benefit to this or should we be confident in the calculation of a single J-integral? Why/why not?

8.7.1 DEFINING CONTOUR INTEGRALS USING CONVENTIONAL FEM VS. XFEM

For conventional FEM, the contour integral region must first be defined. This typically requires specifying the *crack front*, the *crack tip/line*, and the *crack extension direction*. ABAQUS computes the first contour integral (upon request) using all elements inside the crack front and one layer of elements outside the crack front. The software will calculate successive integrals by adding the surrounding single layer of elements to the group of elements that were used to calculate the previous contour integral.

This is shown in Figure 8.9 [3].

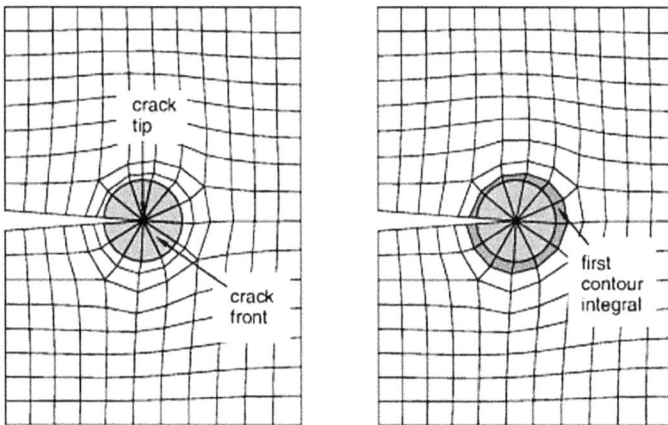

FIGURE 8.9   Contour integral calculation by FEM software (Image source: [3]).

Whether the part instance is a native part, an orphan mesh part, 2D, or 3D will influence the options available for crack front selection. A 2D crack model has edges that are free to move apart while a 3D crack model has faces that are free to move apart. The mesh must match the cracked geometry and a detailed focused local mesh near the crack tip is required.

XFEM does not require the mesh to match the crack geometry; however, enough elements need to be generated around the crack front to enable contour integral calculation. This group of elements within a small radius from the crack front is enriched and becomes involved in the contour integral calculations. In the case of a stationary crack, these elements within a small radius around the crack tip can be used to calculate the crack singularity. This radius is referred to as the enrichment radius.

8.7.2 RESIDUAL STRESSES AND THE J-INTEGRAL

A residual stress field can occur due to thermal effects, service loads that produce plasticity, swelling effects, and certain metal forming processes [4, 5]. When the residual stresses are significant, the standard definition of the J-integral may lead to a path-dependent value [6].

Concept Challenge 8.2

What software efficiencies can be taken advantage of when a crack is located along a line of symmetry?

8.8 INTEGRALS USED IN FEM/XFEM FOR LEFM/EPFM

8.8.1 LINE INTEGRAL

Let f be a function defined on a curve C of finite length, then the *line integral* of f along C is

$$\oint_C f(x,y)\,ds = \lim_{n\to\infty}\sum_1^n f(x_i, y_i)\Delta s_i \tag{8.3}$$

The product for each Δs_i can be viewed as the height of a surface formed in the z-direction. The line integral of a curve along this scalar field will then be equivalent to the area under the curve formed by this surface.

8.8.2 DOMAIN INTEGRAL

The domain integral uses the divergence theorems (2D/3D) to expand the contour integral into an area integral in 2D or a volume integral in 3D, over a finite domain surrounding the crack. The domain integral method has the advantage that contour integral estimates can be obtained with coarse meshes. Green's or the 2D divergence theorem is shown in Eqn. (8.4); the divergence theorem (3D) is shown in Eqn. (8.5).

8.8.3 GREEN'S THEOREM OR 2D DIVERGENCE THEOREM REVIEW

Green's theorem enables the conversion of a line integral to a surface integral. Let C be a positively oriented, piecewise smooth, simple, closed curve and let D be the region enclosed by the curve. If P and Q have continuous first-order partial derivatives on D then,

$$\oint_C P\,dx + Q\,dy = \oint_D \left(\frac{\partial Q}{\partial x} - \frac{\partial P}{\partial y} \right) dA \qquad (8.4)$$

8.8.4 DIVERGENCE THEOREM REVIEW

The divergence theorem enables the conversion of a surface integral to a volume integral.

Let E be a simple solid region and S be the boundary surface of E with positive orientation. Let F be a vector field whose components have continuous first-order partial derivatives, then

$$\iint_S \vec{F} \cdot d\vec{s} = \iiint_E div\left(\vec{F}\right) dV = \iiint_E \left(\frac{\partial F_x}{\partial x} + \frac{\partial F_y}{\partial y} + \frac{\partial F_z}{\partial z} \right) dxdydz \qquad (8.5)$$

8.8.5 INTERACTION INTEGRAL

The concept of the interaction integral (I-integral) was first proposed in 1976 by Stern et al. [7] in order to decouple mixed-mode stress intensity factors (SIFs). The interaction integral method is used in FM software to extract the SIFs from the J-integral results. The J-integral has the physical meaning of the energy release rate G, and the I-integral, I, can be regarded as the mutual energy release rate. The I-integral is path-independent for homogeneous materials and is valid for stress intensity and T-stress (see Section 4.3.2) evaluations of cracks in isotropic and anisotropic media. The I-integral has two limitations [8]:

1. The explicit expression of the auxiliary field must be given in advance.
2. Since the I-integral is based on superposition, it is generally valid for LEFM and finds limited applications in EPFM. For the latter, the inelastic deformation at the crack tip must be small enough that the SIFs still accurately describe the stress field.

8.9 OTHER CONSIDERATIONS

8.9.1 *T-STRESS*

The option to calculate the T-stress may be available in some software packages. Recall Eqn. (2.11) which expresses the crack-tip stress fields in an isotropic elastic material as an infinite power series. The second term does not vanish at the crack tip as it is independent of r. The crack tip stress for a crack under plane strain Mode I loading is shown in Eqn. (8.6), where the T in the second term is called the T-stress

$$\sigma_{ij} = \left(\frac{K_I}{\sqrt{2\pi r}}\right) f_{ij}(\theta) + \begin{bmatrix} T & 0 & 0 \\ 0 & 0 & 0 \\ 0 & 0 & \sqrt{T} \end{bmatrix} \tag{8.6}$$

The T-stress is therefore parallel to the crack faces, *its magnitude can be used as a measure of the usefulness of the J-integral in characterizing the deformation field around the crack*. Though the T-stress is calculated under linear elastic assumptions it is also useful in elastic plastic fracture studies. Several authors [9, 10, 11, 12, 13, 14] have shown that as the T-stress becomes more negative, the tensile stress tri-axiality ahead of the cack tip is reduced, and that this in turn reduces the dominance of the J-integral to describe the Mode I elastic-plastic crack-tip stresses and deformation in plane strain or in 3D.

8.9.2 CONTROLLING THE SINGULARITY AT THE CRACK TIP

For sharp cracks the strain field, ε, at the crack tip will become singular. Some software will allow the inclusion of a singularity within the crack definition. This inclusion may increase the accuracy of the model, particularly for small-strain analysis when geometric nonlinearities are ignored. The material model used in the analysis will affect the crack tip singularity. For example, for the linear elastic case, $\varepsilon \propto r^{-1/2}$, for the perfectly plastic case, $\varepsilon \propto r^{-1}$, for the power-law hardening case,[1] $\varepsilon \propto r^{-\left(\frac{n}{n+1}\right)}$.

8.9.3 SYMMETRY

The existence of symmetry allows the user to take advantage of the reduced computer aided design (CAD) time, computational time, and memory usage. For example, if the crack front is defined on a plane of symmetry, then only half of the structure needs to be modeled.

8.10 FEM EXAMPLE USING XFEM

Example 8.3

Consider the T-beam in Example 7.2 that was subjected to uniform transverse loading and axial loading. Develop a static crack model using XFEM software.

ABAQUS XFEM is used here as an example. In ABAQUS each step (e.g. Apply Mesh, Apply Loads) is provided in a "module" selection window located in the top left corner. Once a step is selected in this window, a menu of icons relevant to that step appear at the top and corresponding icons just below the module, to the left of the dropdown menu shown in Figure 8.10.

Step 1 Draw the beam and crack as separate parts

The beam and crack will not have the same properties; additionally they will interact with each other in a unique manner. Therefore, they are drawn separately, then assembled. They can be drawn using any CAD package and then imported into a computer aided engineering (CAE) package with FM capabilities, or they can be drawn using a single CAD/CAE package that has FM capabilities, such as ABAQUS.

Step 2 Assemble the crack and the beam

This step is shown in Figure 8.11. The "assembly" module is selected to assemble the beam and crack. The crack can then be placed specifically where it needs to be using the "rotate" and "translate" commands. Next, the "create instance" icon is selected; this yields a new window. Both the beam and crack are then selected. Recall that an instance is a replica of a part. There can be several instances for a single part.

Step 3 Define the type of study

To define the type of study, select the "step" module, then select "create step" icon in the main menu. The "static" option can then be selected as shown in Figure 8.12. Selecting "continue" allows you to name the study and provides a time period and increment. The time period refers to the duration of the analysis and the increment describes the frequency of iterations for the numerical solution within the step. Both are relevant when the material response is rate dependent. In a static stress procedure, time-dependent material effects are ignored.

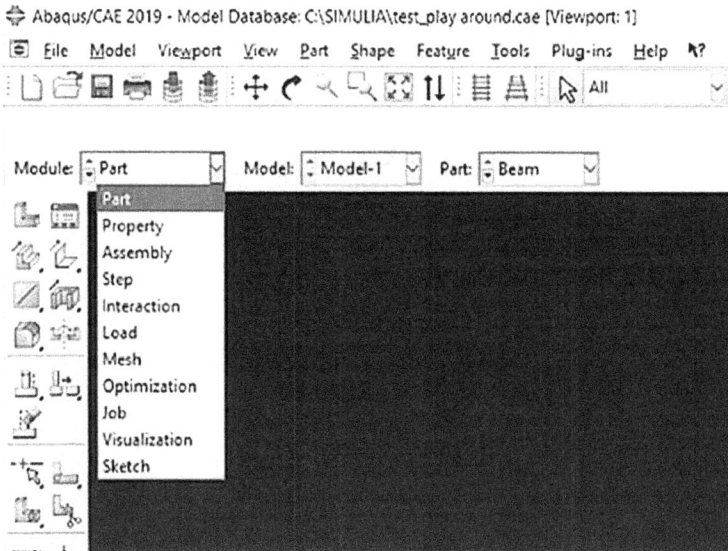

FIGURE 8.10   Module window in ABAQUS (Image source: [3]).

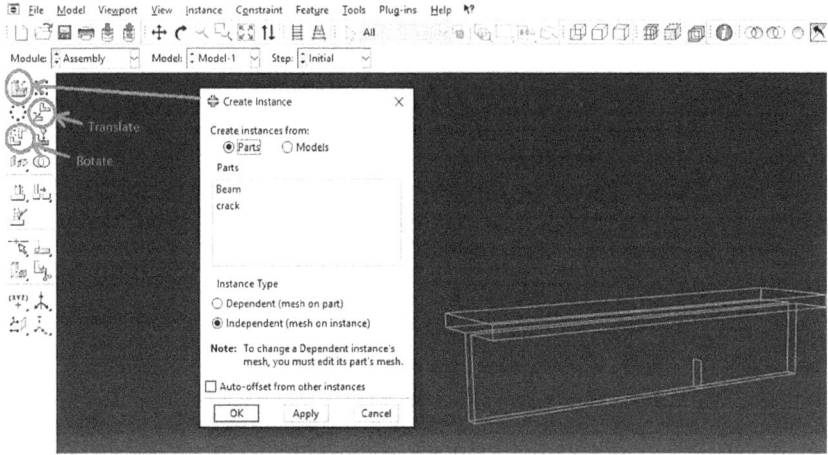

FIGURE 8.11   Assembly of crack and beam (Image source: [3]).

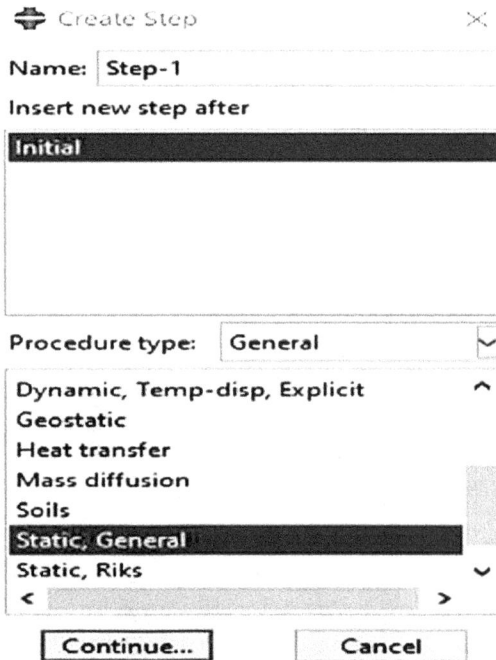

FIGURE 8.12   Step selection in ABAQUS (Image source: [3]).

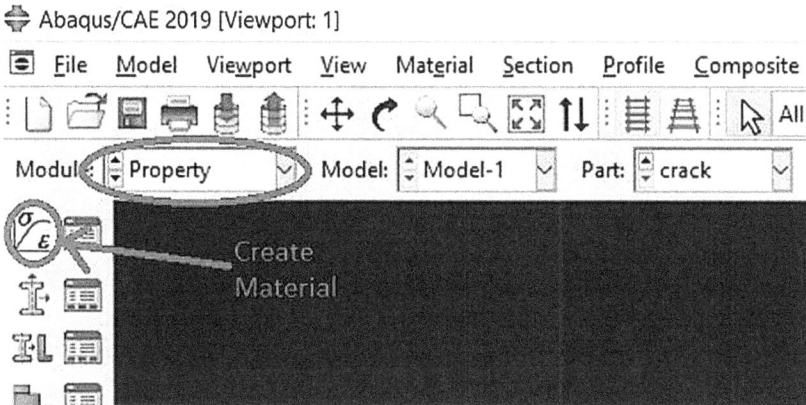

FIGURE 8.13 Defining the material properties in ABAQUS (Image source: [3]).

Step 4 Material definition

Material properties are defined by first selecting the "property" module as shown in Figure 8.13. The "create material" icon can then be selected to define material properties. This action yields a new window which provides options to input mechanical, thermal, or electromagnetic properties. Selecting "Elasticity" then "elastic" in the drop-down menu as shown in Figure 8.14 allows input of the isotropic material's Young's Modulus and Poisson's Ratio. A damage model must also be defined. There are several models available. Choice of model depends on the type of material (metal, polymer, etc.), a history of success (literature) using that damage model for the specific material, the availability of the model parameters or data relevant to the material, and that the model assumptions are compatible with the actual scenario being modeled. In this case a "Damage for traction separation laws" could be chosen and a maximum stress criterion used to predict traction separation.

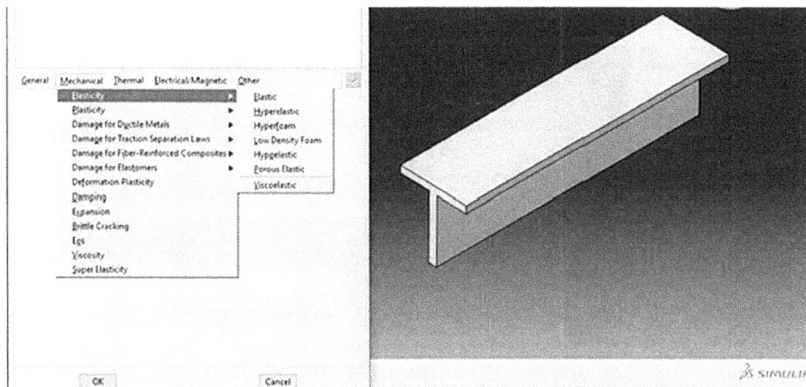

FIGURE 8.14 Entering the material property values in ABAQUS (Image source: [3]).

Step 5 Interaction

As there are two parts, the software will want to know how these parts are intended to interact with each other. The "interaction" module can first be selected as shown in Figure 8.15, the "create interaction" icon (top left) is then chosen, yielding the window with options for a variety of material interactions. In the case, the interaction would be "XFEM crack growth."

Step 6 Loads and boundary conditions

The selection for loads and boundary conditions is shown in Figure 8.16. Loads and boundary conditions are input by first selecting the "Load" module then the "create load" icon (top left). This yields the window shown, where "pressure" would be defined for the top surface, and side surfaces. An option is then provided to enter magnitude and direction (positive or negative). Selecting the "create boundary" condition icon (second from the top left) allows you to input the simply supported displacement conditions at the lower edges at each side of the beam. Loads and boundary conditions for this system are shown in Figure 8.17.

Step 7 Meshing the part or assembly

The part is meshed by selecting the "mesh" module then selecting "Seed Part" (top left) from the main menu. This allows you to define the approximate element size for all edges of a part or part instance. The smaller the element size the greater the computation time, though the greater the accuracy of the result. As with all FE analysis, it's

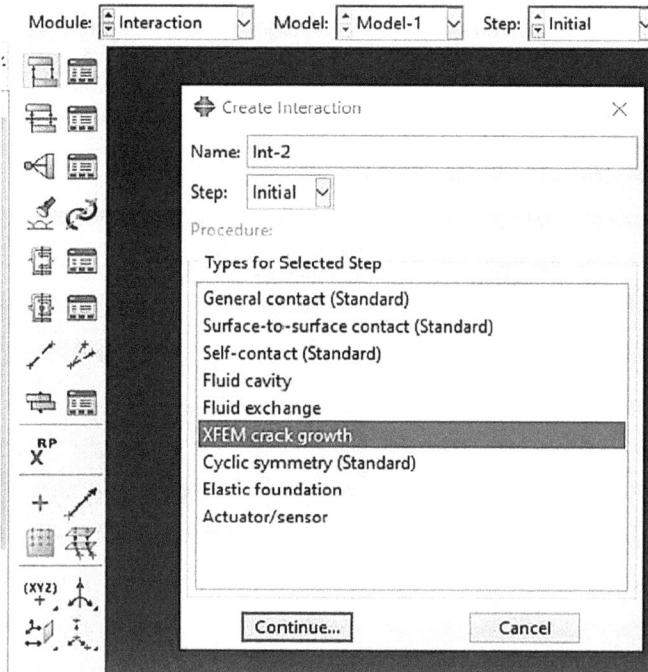

FIGURE 8.15 Interaction property window (Image source: [3]).

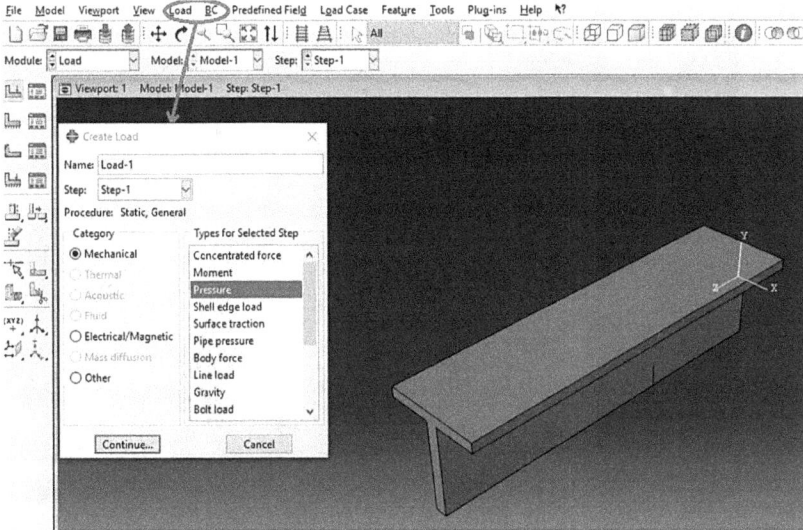

FIGURE 8.16   Entering loads in ABAQUS (Image source: [3]).

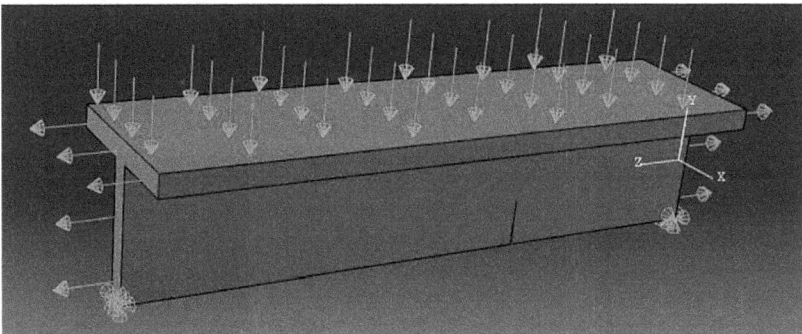

FIGURE 8.17   Axially and transversely loaded, simply supported T-beam in ABAQUS (Image source: [3]).

important to do several runs with decreasing element size until results converge. Next, the "mesh part" icon can be selected; this yields a mesh as shown in Figure 8.18.

Step 8 Performing the analysis

This is done by selecting "Job" in the Module window, then choosing the "create job" icon (top left). This is shown in Figure 8.19. Selecting "Continue" then allows the user to name the job. Selecting the "job manager" icon generates the job manager window shown in Figure 8.20. Selecting "submit" will start the computational process.

Step 9 Viewing results

The job manager window will indicate when the process is complete. You can view output by clicking on the "results" tab. A contour plot of the deformed T-beam is shown

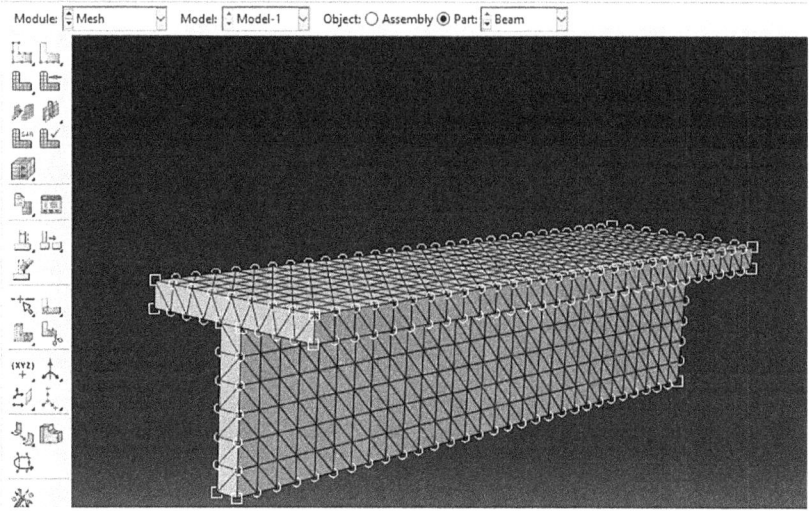

FIGURE 8.18 T-beam meshed in ABAQUS (Image source: [3]).

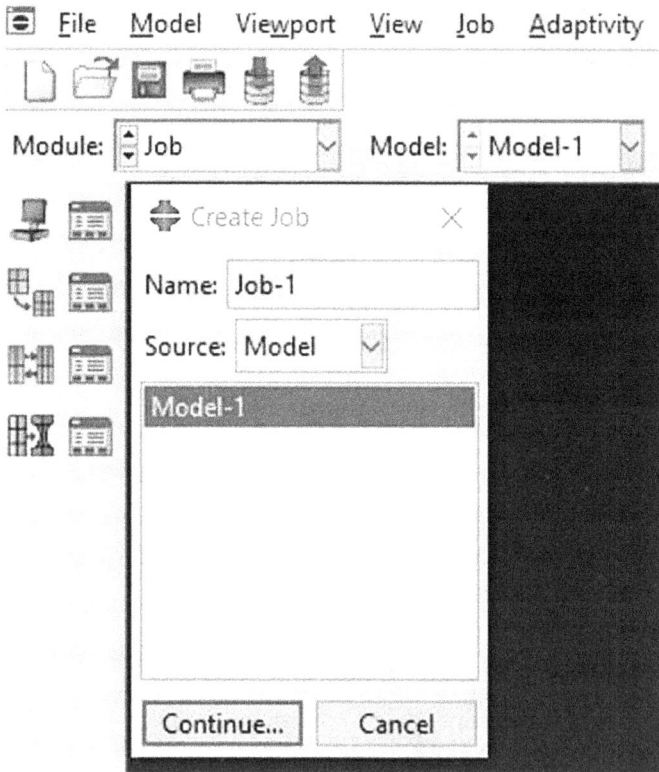

FIGURE 8.19 Creating and initiating a job (Image source: [3]).

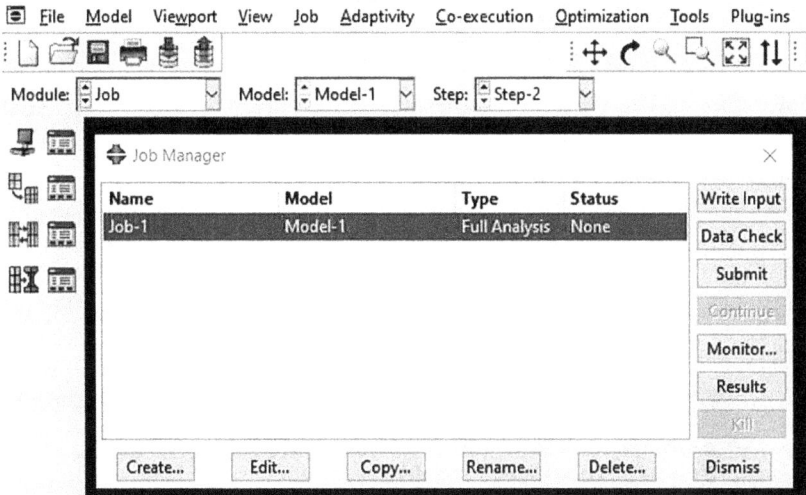

FIGURE 8.20   Job manager window (Image source: [3]).

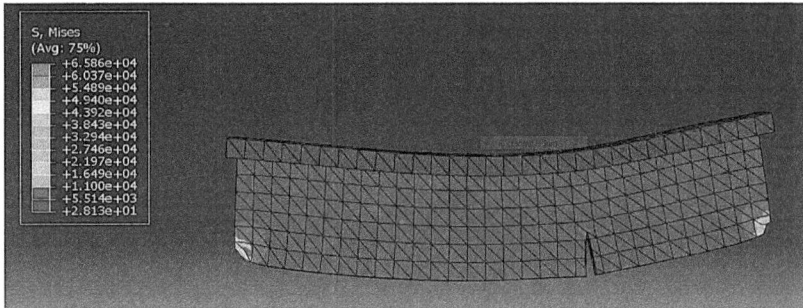

FIGURE 8.21   Contour plot of the deformed T-beam (Image source: [3]).

in Figure 8.21. You are now in the "visualization" module where a variety of outputs such as stresses, strains, and displacements along specific paths or generally overall can be viewed.

8.11   SUMMARY

The ability to determine contour integrals in two or three dimensions gives FEM the ability to determine the onset of cracking. These methods can determine FM parameters such as the J-integral, stress intensity factors for homogeneous materials and interfacial cracks, crack propagation direction, and T-stress. When simulating crack propagation with FEM, cracks debond along user-defined interfaces. XFEM can perform contour integration and does not require a refined mesh at the crack tip. It also does not require remeshing when it simulates the initiation and propagation of a discrete crack along an arbitrary solution dependent path.

QUESTIONS AND PROBLEMS

Write MATLAB or MATHEMATICA code to perform the line integrals for problems 8.1–8.5. Do by hand to verify.

8.1. $\int x^2 \, ds, \, y = \ln x, \left[1, e\right]$

8.2. $\int \dfrac{ds}{y - x}, \, y = \dfrac{x}{2} - 2, \left[0, 4\right]$

8.3. $\int \dfrac{x^3}{y} \, ds, \, y = \dfrac{x^2}{2}, \left[0, 2\right]$

8.4. $\int y e^x \, ds$ over a line segment from [0,2] to [3, 4]
8.5. $\int z x^2 \, ds$ over a line segment from [0,6, −1] to [4, 1, 5]
8.6.   What is the difference between a native part and an orphaned mesh? Explain how a crack is defined for each case.
8.7.   Explain what is meant by "convergence" and how it is applied regarding predictions for an FEM crack model.
8.8.   If the part is symmetrical, what is the benefit of only modeling one-half of the part that is consistent with the symmetry plane?
8.9.   Visit the National Transportation and Safety Board website. Navigate to the reports link. Research a case where a single crack instability due to fatigue or overload was involved. Write the steps that would be involved if that crack were modeled using FEM. What would your boundary conditions and loading be?
8.10. What is meant by path independence? The presence of residual stresses disallows path independence for the J-integral. Research and explain on a general level what FEM software may incorporate to manage this issue.
8.11. Why bother with the J-integral as a measure of energy release rate? Could we use FEM to solve the boundary value problem, obtain the stress/strain field, and integrate to obtain the elastic energy at different crack incremental lengths to do this? Why? Why not?
8.12. T-stress is determined based on linear elastic fracture mechanics. Why might T-stress be relevant in an EPFM problem where the J-integral is calculated?

NOTE

1 Power-law hardening has the form $\dfrac{\overline{\varepsilon}}{\varepsilon_0} = \alpha \left(\dfrac{\overline{\sigma}}{\sigma_0} \right)^n$ where $\overline{\varepsilon}$ is the equivalent total strain, ε_0 is a reference strain, $\overline{\sigma}$ is the Mises stress, σ_0 is the initial yield stress, n is the power-law hardening exponent (typically in the range of 3 to 8), and α is a material constant (typically in the range 0.5 to 1.0) .

REFERENCES

[1] Dassault Systemes Siumlia Corp, "ABAQUS/CAE 2019 User Manual," Dassault Sytemes Simulia Corp., Johnston, RI, 2019.

[2] Ansys Inc. "Ansys SMART Fracture White Paper," Anys Inc, Canonsburg, PA, 2020.

[3] Dassault Systemes, ABAQUS CAE User Manual: Using contour integrals to model fracture mechanics, Vélizy-Villacoublay: Dassault Systemes, 2019.

[4] X. Rena, Z. Zhanga and B. Nyhus, "Effect of residual stresses on the crack-tip constraint in a modified boundary layer model," *International Journal of Solids and Structures*, vol. 46, no. 13, pp. 2629–2641, 2009.

[5] C. Coates and V. Vakharia, "The Effect of Imperfections on Crack growth Rates in Cold Worked and Interference Fit holes," *Scientific Technical Review*, vol. 69, no. 3, pp. 35–40, 2019.

[6] Y. Lei, N. O'Dowd and G. Webster, "Fracture mechanics analysis of a crack in a residual stress field," *International Journal of Fracture*, vol. 106, pp. 195–216, 2000.

[7] M. Stern, E.B. Becker and R. S. Dunham, "A Contour Integral Computation of Mixed-mode Stress intensity Factors," *International Journal of Fracture*, vol. 12, p. 359–368, 1976.

[8] H. Yu and M. Kun, "Interaction integral method for computation of crack parameters," *Engineering Fracture Mechanics*, vol. 249, 2021. doi:10.1016/j.engfracmech.2021.107722.

[9] Y. Y. Wang, "A Two-Parameter Characterization of Elastic-Plastic Crack Tip Fields and Application to Cleavage Fracture," PhD thesis, Department of Mechanical Engneering, MIT, Cambridge, 1991.

[10] D. M. Parks, "Advances in Characterization of Elastic-Plastic Crack-Tip Fields," in *Topics in Fracture and Fatigue*, Springer Verlag, 1992.

[11] Z. Z. Du and J. W. Hancock, "The Effect of Non-Singular Stresses on Crack-Tip Constraint," *Journal of the Mechanics and Physics of Solids*, vol. 39, pp. 555–567, 1991.

[12] A. M. Al-Ani and J. Hancock, "J-Dominance of Short Cracks in Tension and Bending," *Journal of the Mechanics and Physics of Solids*, vol. 39, pp. 23–43, 1991.

[13] C. Betegón and J. Hancock, "Two-Parameter Characterization of Elastic-Plastic Crack-Tip Fields," *Journal of Applied Mechanics*, vol. 58, pp. 104–110, 1991.

[14] B. A. Bilby, G. Goldthorpe and I. Howard, "A Finite Element Investigation of the Effect of Specimen Geometry on the Fields of Stress and Strain at the Tip of Stationary Cracks," in *Size Effects in Fracture*, London, Mechanical Engineering Publications Ltd for the Institution of Mechanical Engineers, 1986, pp. 37–46.

Appendix A

A.1 MECHANICS OF MATERIALS REVIEW

Stress, σ, is defined as the force carried by a component per unit cross-sectional area. The force, $\vec{F}$, may be perpendicular (normal) or parallel (shear) to the area. The stress definition is:

$$\sigma = \frac{\vec{F}}{A} \tag{A.1}$$

Strain, ε, is the extension, ΔL, per unit length, L, due to a normal force, or the relative in plane displacement of parallel layers per unit separation distance between the layers due to a shear force. The strain definition is:

$$\varepsilon = \frac{\Delta L}{L} \tag{A.2}$$

The application of a normal stress will result in a normal strain response, while the application of a shear stress will result in a shear strain response. Figure A.1 depicts the stress-strain response of a coupon (homogeneous isotropic material) subjected to tensile loading. The loading is tensile, therefore the stress on the cross-sectional area is a normal stress; the resulting strain is therefore a normal strain. The stress applied is initially linearly related to the strain response until the stress achieves a certain magnitude, referred to as the yield strength. Up to this value, the member will return to its original length if unloading occurs. The region where this behavior occurs is called the *elastic* region and the strain in this region is the *elastic* strain. Beyond the yield strength the response is no longer linear, the member will not return to its original length if unloaded. This latter region is the plastic region, the strain that occurs here is called the *plastic* strain.

In the selection of appropriate material for engineering applications, there is always a trade-off regarding the relevant material properties. Material properties most frequently used in engineering analysis include:

- Ultimate tensile strength: when subjected to load, the maximum stress the material can withstand;
- Yield strength: when subjected to load, the maximum stress the material can withstand before transitioning from elastic to plastic (i.e. some permanent deformation takes place);
- Fracture strength: when subjected to load, the stress value at which the material breaks;
- Young's modulus: the ratio of the material's stress to its strain in the linear elastic region;

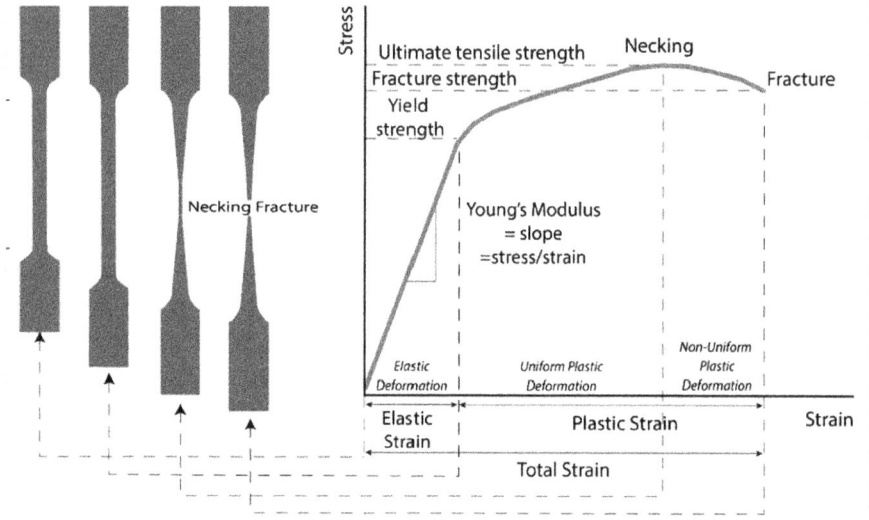

FIGURE A.1 Stress vs. strain response for tensile loaded specimen.

- Shear modulus: the ratio of the material's shear stress to its shear strain in the elastic region;
- Ductility: is a measure of the amount of plastic deformation that occurs before fracture;
- Toughness: refers to the amount of energy a material without flaws can absorb before breaking;
- Impact strength: the amount of energy that a material can withstand before breaking when subjected to a load that is suddenly applied over a short period of time.

Not all materials exhibit initial linear elastic behavior; however, most materials currently used in engineering applications exhibit some initial linear elasticity. Materials may also exhibit an initial nonlinear elastic or inelastic response to applied loads.

A.2 STRESS TRANSFORMATION, PRINCIPAL STRESS, STRAIN

For plane stress, the stresses on one plane are zero; let's choose the z-plane, that is the plane perpendicular to the z-axis, then $\sigma_{zz} = \sigma_{xz} = \sigma_{yz} = 0$. The 3D element in plane stress can therefore be represented as a 2D element in the xy plane, as shown in Figure A.2(a). The stresses acting on the element rotated through an angle θ about the z-axis to a new coordinate system $x'y'$ can be expressed in terms of the stresses acting on the element in the xy coordinate system using the equations of equilibrium. The rotated element is shown in Figure A.2(b). These relationships are described by:

$$\sigma_{x'x'} = \frac{\sigma_{xx} + \sigma_{yy}}{2} + \frac{\sigma_{xx} - \sigma_{yy}}{2}\cos 2\theta + \sigma_{xy}\sin 2\theta \qquad (A.3)$$

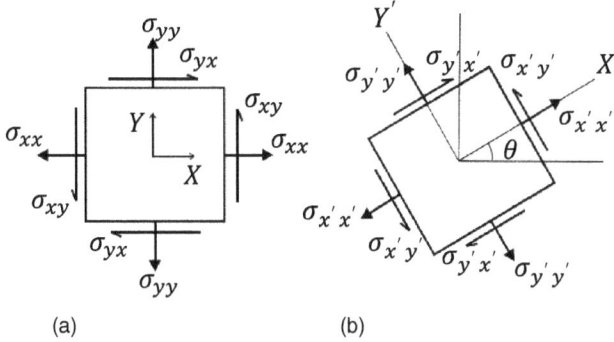

FIGURE A.2   (a) Plane stress (z-plane is stress free). (b) Stresses transformed to another coordinate system ($x'y'$).

$$\sigma_{x'y'} = \frac{\sigma_{xx} - \sigma_{yy}}{2} \sin 2\theta + \sigma_{xy} \cos 2\theta \qquad (A.4)$$

The normal stress acting on the face perpendicular to the y', i.e. $\sigma_{y'y'}$, may found by substituting $\theta + 90^0$ for θ in Eqn. (A.3), this yields

$$\sigma_{y'y'} = \frac{\sigma_{xx} + \sigma_{yy}}{2} - \frac{\sigma_{xx} - \sigma_{yy}}{2} \cos 2\theta - \sigma_{xy} \sin 2\theta \qquad (A.5)$$

Since Eqn. (A.3)–(A.5) transform the stress components from one coordinate system to another, they are referred to as the *transformation equations for plane stress*.

Adding Eqn. (A.3) and (A.5), we obtain

$$\sigma_{x'x'} + \sigma_{y'y'} = \sigma_{xx} + \sigma_{yy} \qquad (A.6)$$

Therefore, as a plane stress element is rotated, the sum of the normal stresses acting on perpendicular faces remains constant and independent of the rotation angle, θ.

The value of θ that yields extremum valuesfor the transformed stresses may be found by differentiating Eqn. (A.3) and (A.4) with respect to θ and setting them to zero. Performing this action on Eqn. (A.3) and denoting the value of θ that yields extremum normal stress values as θ_p,we obtain

$$\tan 2\theta_p = \frac{2\sigma_{xy}}{\sigma_{xx} - \sigma_{yy}} \qquad (A.7)$$

When the state of stress for one orientation is known, i.e. σ_{xx}, σ_{xy}, and σ_{xy}, solving Eqn. (A.7) for θ_p will yield two values of θ_p that differ by $90°$. These values define the orientation of the element for which one of the normal stresses is a maximum while the other is a minimum. These extremum stress values are called *principal stresses* and the planes on which they act are called *principal planes*.

If we set $\sigma_{x'y'}$ to zero in the transformed shear stress Eqn. (A.4), we arrive at the same expression as shown in Eqn. (A.7), therefore *the shear stresses are zero on the principal planes.*

Differentiating Eqn. (A.4) with respect to θ and setting to zero, yields

$$\tan 2\theta_s = -\frac{\sigma_{xx} - \sigma_{yy}}{2\sigma_{xy}} \tag{A.8}$$

The subscript s indicates that the angle θ_s defines the orientation of the planes of maximum positive and maximum negative shear stresses.

On inspection of Eqn. (A.7) and (A.8), we see that

$$\tan 2\theta_s = -\frac{1}{\tan 2\theta_p} = -\cot 2\theta_p \tag{A.9}$$

$$\frac{\sin 2\theta_s}{\cos 2\theta_s} = -\frac{\cos 2\theta_p}{\sin 2\theta_p}$$

$$\cos 2\theta_s \cos 2\theta_p + \sin 2\theta_s \sin 2\theta_p = 0$$

Recall the identity $\cos(A - B) = \cos A \cos B + \sin A \sin B$, therefore

$$\cos\left(2\theta_s - 2\theta_p\right) = 0$$
$$2\theta_s - 2\theta_p = \pm 90°$$
$$\theta_s = \theta_p \pm 45°$$

This result yields an important insight that *the planes of maximum shear occur at 45° to the principal planes.*

Concept Challenge

Are the transformation equations for plane stress applicable to nonlinear elastic materials? What about inelastic materials? Why? Why not?

Appendix B

TABLE B.1
Finite Size Correction for Common Geometries ($\alpha = a/W$)

Case	Diagram	SIF Expression
Through crack in a finite plate in tension		$Y = \sqrt{\sec\left(\pi\alpha/2\right)}$
Edge crack in finite width plate in tension		$Y = \dfrac{1.12 + \alpha\left(\left(2.91\alpha - 0.64\right)\right)}{1 - 0.93\alpha}$
Edge crack in finite width plate in pure bending		$K_I = \left(6M/bW^2\right)Y\sqrt{\pi a}$ $Y = \dfrac{1.12 + \alpha\left(\left(2.62\alpha - 1.59\right)\right)}{1 - 0.7\alpha}$
Circumferential crack in circular rod in tension		$Y = \dfrac{1.12 + \alpha\left(\left(1.30\alpha - 0.88\right)\right)}{1 - 0.92\alpha}$

Appendix C

TABLE C.1

$f\left(\dfrac{a}{W}\right)$ Solutions for Common Test Specimens Where $K_I = \dfrac{P}{B\sqrt{W}}\, f\left(\dfrac{a}{W}\right)$

and B is the Specimen Thickness

Geometry	$f\left(\dfrac{a}{W}\right)$

Compact specimen

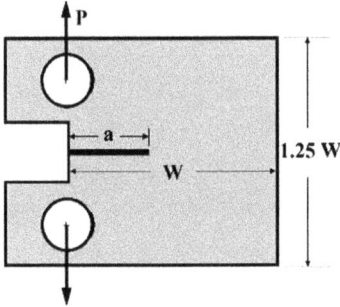

$$\dfrac{2+(a/W)}{\left(1-(a/W)\right)^{3/2}}$$

$$\left[0.886 + 4.64\left(\dfrac{a}{W}\right) - 13.32\left(\dfrac{a}{W}\right)^2 \right.$$
$$\left. + 14.72\left(\dfrac{a}{W}\right)^3 - 5.60\left(\dfrac{a}{W}\right)^4\right]$$

Single-edge notched bend (SE(B))

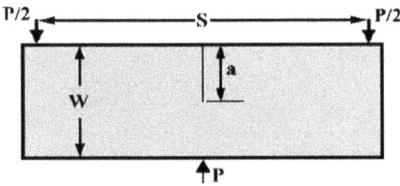

$$\dfrac{3(S/W)\sqrt{a/W}}{2\left(1+2(a/W)\right)\left(1-(a/W)\right)^{3/2}}$$

$$\left[1.99 - \dfrac{a}{W}\left(1-\dfrac{a}{W}\right)\right.$$
$$\left.\left\{2.15 - 3.93\left(\dfrac{a}{W}\right) + 2.7\left(\dfrac{a}{W}\right)^2\right\}\right]$$

Single-edge notched tension (SENT)

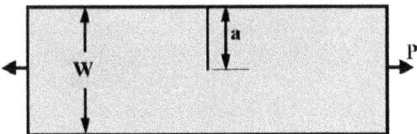

$$\dfrac{\sqrt{2\tan\left(\pi a/2W\right)}}{\cos\left(\pi a/2W\right)}$$

$$\left[0.752 + 2.02\left(\dfrac{a}{W}\right) + 0.37\left(1 - \sin\left(\dfrac{\pi a}{2W}\right)\right)^3\right]$$

(Continued)

225

TABLE C.1 (CONTINUED)

Double-edge notched tension (DENT)

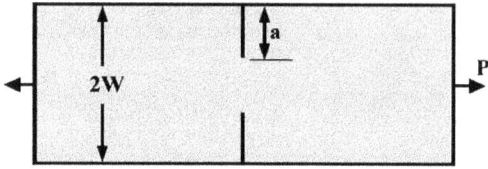

$$\frac{\sqrt{(\pi a/2W)}}{\sqrt{1-(a/W)}}$$

$$\left[1.122-0.561\left(\frac{a}{W}\right)-0.205\left(\frac{a}{W}\right)^2 +0.471\left(\frac{a}{W}\right)^3+0.190\left(\frac{a}{W}\right)^4\right]$$

Center-cracked tension (CCT)

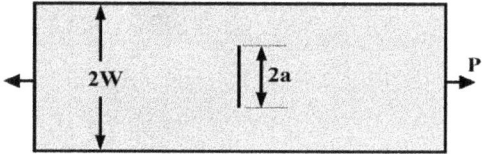

$$\sqrt{\frac{\pi a}{4W}\sec\frac{\pi a}{2W}}$$

$$\left[1-0.025\left(\frac{a}{W}\right)^2+0.06\left(\frac{a}{W}\right)^4\right]$$

H. Tada, P.C. Paris, and G.R. Irwin *The Stress Analysis of Cracks Handbook*, 2nd ed.. St. Louis, Paris Productions, 1985.

Index

227